NUCLEAR SERVITUDE

Subcontracting and Health in the French Civil Nuclear Industry

Annie Thébaud-Mony

Director of Research
National Institute for Health & Medical Research
(INSERM) France

Translated from the French by Amy Jacobs-Colas

Work, Health, and Environment Series
***Series Editors*: Charles Levenstein, Robert Forrant, and John Wooding**

Baywood Publishing Company, Inc.
AMITYVILLE, NEW YORK

Baywood Publishing Company, Inc.
26 Austin Avenue
PO Box 337
Amityville, NY 11701
(800) 638-7819
E-mail: baywood@baywood.com
Web site: baywood.com

Library of Congress Catalog Number: 2009043025
ISBN: 978-0-89503-380-2 (cloth : alk. paper)
ISBN: 978-0-89503-424-3 (epub)
ISBN: 978-0-89503-450-2 (epdf)
http://dx.doi.org/10.2190/NSS

Library of Congress Cataloging-in-Publication Data

Thébaud-Mony, Annie, 1944-
[Industrie nucléaire. English]
Nuclear servitude : subcontracting and health in the French civil nuclear industry / Annie Thébaud-Mony ; translated from the French by Amy Jacobs-Colas.
p ; cm. -- (Work, health, and environment series)
Includes bibliographical references and index.
ISBN 978-0-89503-380-2 (cloth : alk. paper)
1. Nuclear industry--Health aspects--France. 2. Nuclear power plants--Employees--Health--Health and hygiene--France. 3. Subcontracting--Health aspects--France. I. Title. II. Series: Work, health, and environment series.
[DNLM: 1. radiation Injuries--prevention & control--France. 2. Contract Services--standards--France. 3. Nuclear Reactors--standards--France. 4. Occupational Exposure--prevention & control--France. WN 610 T375i 2020a]
HD7269.A62F813 2010
331.25'60944--dc22

2009043025

Dedication

I dedicate the book to my grandchildren,
Maël, Marie, Anne, Lucah, and Kezia:
may the world in which they grow up be one that works
attentively and effectively to safeguard life, and
may this book have some effect to that end.

Nuclear Servitude: Subcontracting and Health in the French Civil Nuclear Industry was originally published in French as *L'Industrie nucléaire: sous-traitance et servitude,* copyright 2000 Editions INSERM/EDK, Paris.

Assistance for the translation was provided by the Mission Interministérielle pour la Recherche (MIRe) and the Comités Mixtes à la Production EDF.

Table of Contents

Acknowledgments

This work owes much to the many groups and individuals who helped me undertake it, pursue it, and bring it to completion:

- the nuclear industry "outside" workers who spoke to me of what they do, giving me to understand their work and the meaning they attribute to it as a public service dedicated to nuclear safety;
- all the persons who worked with me at successive stages of the research: Catherine Rondeau du Noyer, Catherine Brousselle, Dominique Huez and the Région Centre occupational doctors' group;
- Michel Lallier and the Confédération Générale du Travail (CGT) union coordination committee for nuclear workers at the Chinon plant site;
- Catherine Pozzo di Borgo, who made the documentary "Arrêt de tranche: les trimardeurs du nucléaire" ["Unit shutdown: the vagabonds of the nuclear industry"];
- the two INSERM research teams I was successively part of while doing the research: the ISIS group [Inégalités sociales, Industrialisation et santé] of the Unit 292 (Kremlin-Bicêtre) and the Centre de Recherche sur les Enjeux Contemporains en Santé Publique at the University of Paris 13 (Bobigny);
- the networks of associations that helped me understand the implications of this study for occupational and public health policy;
- colleagues and friends who generously reread drafts of the work and discussed its hypotheses with me;
- the French Ministry of Labor and the Réseau National de Santé Publique [National public health network] for their institutional and financial support;
- Suzy Mouchet and the Editions INSERM.

For the English edition, I am grateful to my translator Amy Jacobs-Colas, who helped me clarify a number of points for an audience unfamiliar with French institutions and the specific situation studied.

Lastly, I thank my family for their presence and affection, which ensured me the serenity I needed to write.

Preface

In the Baywood series, *Work, Health, and Environment,* the conjunction of topics is deliberate and critical. We begin at the point of production—even in the volumes that address environmental issues—because that is where things get made, workers labor, and raw materials are fashioned into products. It is also where things get stored or moved, analyzed or processed, computerized or tracked. In addition, it is where the folks who do the work are exposed to a growing litany of harmful things or are placed in harm's way. The focus on the point of production provides a framework for understanding the contradictions of the modern political economy.

Despite claims to a post-industrial society, work remains essential to all our lives. While work brings income and meaning, it also brings danger and threats to health. The point of production, where goods and services are produced, is also the source of environmental contamination and pollution. Thus, work, health, and environment are intimately linked.

Work organizations, systems of management, indeed the idea of the "market" itself, have a profound impact on the handling of hazardous materials and processes. The existence or absence of decent and safe work is a key determinant of the health of the individual and the community: what we make goes into the world, sometimes improving it, but too often threatening the environment and the lives of people across the globe.

We began this series to bring together some of the best thinking and research from academics, activists, and professionals, all of whom understand the intersection between work and health and environmental degradation, and all of whom think something should be done to improve the situation.

The works in this series stress the political and social struggles surrounding the fight for safer work and protection of the environment, and the local and global struggle for a sustainable world. The books document the horrors of cotton dust, the appalling and dangerous conditions in the oil industry, the unsafe ways in which toys and sneakers are produced, the struggles to link unions and communities to fight corporate pollution, and the dangers posed by the petrochemical industry, both here and abroad. The books speak directly about the contradictory effects of the point of production for the health of workers, community, and the environment. In all these works, the authors keep the politics front and foremost. What has emerged, as this series has grown, is a body of scholarship uniquely focused and highly integrated around themes and problems absolutely critical to our own and our children's future.

Preface to the English Edition

> For if it is scandalous to die, it is even more scandalous when the traces of the dead degenerate into the death of all traces.
>
> Louis-Vincent Thomas (1991)

Considered at the international level, the civil nuclear industry has been with us for slightly over half a century. In France, with its 59 operating reactors, the industry is now the first source of electricity: 78% of all electric kilowatt-hours produced in France are of nuclear origin.[1] French nuclear plants are operated by two major categories of workers: permanent employees of *Electricité de France* (EDF), particularly reactor operators; and maintenance workers who perform reactor inspection, upkeep, and repair. Maintenance is essential to nuclear plant operation and safety. And it includes the majority of work activities involving exposure to ionizing radiation. In France as in most of the other European Union countries with nuclear power plants, maintenance work is subcontracted to outside companies.

The hypothesis at the origin of the research presented in this book is that the choice to subcontract nuclear plant maintenance was based on both structural developments in employment, work, and occupational health over the last 30 years and perhaps above all on the contradictions specific to the nuclear industry. Employees of subcontracted or "outside" companies, whom in France are often designated DATRs—"directly assigned to tasks exposed to radiation"[2]—take

[1] "There are now in the world some 436 commercial nuclear power reactors operating in 30 countries, with over 372,000 MWe of total capacity. France comes second behind Lithuania, 80% of whose electricity is produced by the nuclear industry. Nuclear plants supply 16% of the world's electricity, as base-load power, and their efficiency is increasing. Fifty-six countries operate a total of 250 research reactors and a further 220 reactors power ships and submarines" (see http://www.world-nuclear.org/info/reactors.htm).

[2] According to the *Code du Travail* [official compilation of French labor laws and regulations], "DATR" refers to all workers exposed to radiation. DATRs must wear a device that measures the degree of their exposure. This applies not only to all exposed personnel implicated in the nuclear cycle (production of electricity, extraction and processing of nuclear fuel, reprocessing and managing of nuclear waste) but also to workers in other industries that use radioactive sources and to medical personnel working in radiology or radiotherapy.

80% of the total annual occupational exposure to ionizing radiation in French nuclear plants. This study begins with some of these workers' accounts of their experience, then presents and analyzes the contradictions that the nuclear industry has to confront, caught as it is between economic constraints and those of radiation protection regulations; between worker health and safety and the safety of nuclear facilities for surrounding populations, the environment and the world at large. It was conducted in response to the alert sounded by unionists and occupational health specialists at the Chinon nuclear plant, on the banks of the Loire River in central France. They were concerned about working and health conditions for workers whom journalists call the "intermittents" or "nomads" or "vagabonds" of the nuclear industry. It takes the form of a longitudinal qualitative survey of DATR workers pursued over ten years, from 1988 to 1997; the results were published in France in 2000 under the title *L'industrie nucléaire: sous-traitance et servitude*. It shows how the legitimacy of subcontracting nuclear plant maintenance work was constructed and the consequences of that way of proceeding for workers' health and overall plant safety.

The nuclear industry, one of the most recent progeny of the industrial era, is also an industry where risks to health in the short, medium, and long terms seem both the most dreaded and the most controversial. What are the health effects of low doses of ionizing radiation? What is the impact on humans of contamination by radioactive particles, particularly internal contamination? The 20th anniversary of the Chernobyl nuclear disaster has brought into the public eye and the public space an intense controversy on these questions, not only in the countries most directly affected by the disaster (Belarus, Ukraine, and Russia) but also elsewhere in Europe, and particularly in France. The catastrophe was also a blow to the general social representation of civil nuclear power: the unthinkable had occurred.[3] And it substantiated the meaning of that "unthinkable." The title of an article marking the 20th anniversary of the disaster by Svetlana Alexievich, the Belarusan author of *Voices from Chernobyl,* takes the form of a question: *"Chernobyl: our past or our future?"* (see *Le Monde* of April 25, 2006).

The debate on the effects of low-dose radiation, confined until the first decade of the 21st century to narrow circles in which a few independent scientists[4] stood opposed to national and international specialized nuclear energy agencies (among others the International Commission on Radiation Protection or ICRP and the International Atomic Energy Agency or IAEA), entered the public arena

[3]In fact, studies by the Brookhaven physicists done as early as 1957 had provided a full account of the possible consequences of an accident involving a 500-megawatt nuclear reactor (WASH 740). But their work was not made public.

[4]The independent scientists include Alice Stewart et al. (1980, 1982), Thomas Mancuso (1977, 1980, 1981) and, in France, Bella and Roger Belbeoch who have been publishing studies on the subject since 1978 (see in particular Belbeoch, 1988, 1993). Official bibliographies on the problem have consistently ignored these publications.

by way of three recent events, events which made it clear that the reference model for effects of ionizing radiation on health was beginning to be called into question.

The first of these, in 1999, was the sudden interruption of the research work of Yuri Bendazhevsky, and his arrest and imprisonment for "activities threatening state security." Bendazhevsky was studying the multiple effects of Chernobyl radioactive contamination on the health of Belarusan children (Bendazhevsky, 2005). The event provoked an international movement of support and call for his release, which paradoxically gave world visibility to his research results—results that contradict the optimistic appraisals of the health consequences of the Chernobyl accident published by international organizations.[5]

The second event was the social movement launched by the *Nuclear Test Veterans Association*, known in France as the *Association des Vétérans des Essais Nucléaires* (AVEN), who, since 2000, in several countries have been publicizing the issue of the health consequences of nuclear testing fallout, namely through court actions (Barillot, 2003).[6]

The third event was the establishment of the *European Committee on Radiation Risks* (ECRR), on the initiative of a European Parliament group. The ECRR brings together independent scientists without any contractual relations to the nuclear industry or to national and international groups specialized in atomic energy and the risks thereof. In 2003, the committee published the results of a collective expert study of the effects of low-dose radiation (ECRR, 2003); then in 2006 a synthesis of works on the health effects of the Chernobyl accident, most of which had been published only in Russian (ECRR, 2006). The ECRR studies have had the valuable effect of broadening discussion, which up till very recently in France had been confined to the question of whether or not there exists a toxicity threshold for external irradiation; they have pointed up the need for scientific studies of internal irradiation.

None of the studies either published by Bendazhevsky on the multiple effects of Chernobyl radioactive contamination or reviewed by the ECRR are referenced in the official scientific literature. The last example is an article of the International Agency for Cancer Research (IARC) presenting a synthesis of the studies on the *Epidemiology of the after Chernobyl* published in the French Bulletin du Cancer (Kesminiene & Cardis, 2007).

[5]See the site of the International Atomic Energy Agency: http://www.iaea.org/NewsCenter/Focus/index.html.

[6]An AVEN study of 1500 French veterans shows the high prevalence of cancer (34.5% of respondents suffer from one or several types of cancer and 41% have died, 2/3 of them before the age of 60) and others pathologies. Effects on descendants have also been observed: high prevalence of miscarriages among veterans' spouses, congenital anomalies in their children. Also noted are problems of sterility due to sperm anomalies among veterans themselves (AVEN, 2006).

However, the question of work involving exposure to radiation and the health of the workers who do it has not really been part of the debate. In fact, our improved scientific understanding of the effects of low-dose radiation calls into question the very foundations for current radiation protection norms, particularly for workers chronically exposed to radiation in this way.

This preface reviews recent studies on the destabilizing of employment and its effects on health in the workplace, and recent biological and epidemiological studies of health effects of low level doses of ionizing radiation, making clear the continuing relevance of the questions and issued raised in the book.

1. THE EFFECTS OF IONIZING RADIATION: QUESTIONING THE CURRENT MODEL

On nuclear industry sites in all countries, workers undergo chronic occupational exposure to what is called "artificial" radiation; that is, radiation generated by the industrial production process itself. The fact is that there is no clear consensus on the health consequences of this type of exposure.

The controversy, discussed in detail in chapter 6, is not new. However, it has reached a major turning point today. Until now, the authoritative reference for understanding the effects of ionizing radiation, as defined by the *International Commission on Radiological Protection* or ICRP, has been based on the results of an epidemiological study of survivors of the atomic bombs dropped on Hiroshima and Nagasaki. The theoretical foundations for that model were the following three postulates: the effects of low-dose radiation on health are directly relevant for and may be extrapolated from the effects of high-dose radiation; the only long-term effect of radiation is cancer; there is no toxicity threshold but instead a "linear no-threshold" relation between exposure to ionizing radiation and the onset of cancer (see chapter 5).

The ICRP explains that it is not possible to set radiation exposure limiting norms exclusively in reference to scientifically established knowledge about health. The norms developed out of a compromise adopted by the ICRP in which three ranges of risk, termed respectively "acceptable," "tolerable," and "intolerable" are defined. The ICRP definition of an "acceptable" risk has itself changed over time but is still calculated in terms of a "costs-benefits" relation determined by experts wherein the "price of human life" is put into the balance with the benefits expected from the nuclear industry. ICRP recommendations specify that the limits adopted are not "non-effect" limits (ICRP60, 1990). Still, as discussed in chapter 5, the type of practical organization of work used to manage irradiation is aimed first and foremost at keeping exposure levels below a certain threshold, particularly occupational exposure levels. The dominant representation of these norms among exposed populations is that complying with the dose maximums imposed by those norms guarantees protection of exposed persons' health. This representation was strengthened in France by a public debate initiated by the French academies of science and medicine in the 1990s on the question of whether or not there is a toxicity threshold for ionizing radiation.[7] Citing cellular biology experiment data, the French academicians were adamantly opposed to having these norms lowered. They have constantly contested the notion of a no-threshold linear relation between ionizing radiation and cancer (Tubiana & Aurengo, 2005, 2006), despite the fact that researchers whose work has never been challenged at the international level reject the French academicians' hypotheses (Brenner, 2006).

The Recommendations of the European Committee on Radiation Risks

In its 2003 report on the effects of low-dose radiation, the ECRR not only cites cellular biology studies but also an entire series of experimental, epidemiological, and observation studies, including studies of the health consequences of the Chernobyl disaster done by Russian and Belarusan researchers, works that call into question the validity of the ICRP-established model based on results of the cohort study of survivors of the Hiroshima and Nagasaki bombings. On the basis of this critical review of international literature, the ECRR proposed a new interpretive model wherein each type of exposure is considered in terms of the "trace structure" in space and time of radiation in the cells. With this biological model, the idea that the effects of low-dose radiation on the human organism may be extrapolated from those of high-dose radiation is rejected. The ECRR experts hypothesize different effects by type of irradiation. Specifically, they assert that internal irradiation (that is, through inhalation or ingestion of radioactive dust, which then becomes a radioactive source within the organism) has different effects than the more readily measurable ones of external irradiation. They support this hypothesis with references to several phenomena necessarily implied in differentiating radiation effects: the extremely diverse processes leading to genetic mutation and other forms of biological damage; the variable sensitivity to ionizing radiation of different organs, tissues, and cells; the particular characteristics of the radiation received (α, β, λ, neutrons), itself variable, and dose received (strong or weak discharge, whether dose was received intermittently or not, fractionation).

Whereas for the ICRP cancer is implicitly understood to be the effect of a single event, the ECRR, citing fundamental scientific knowledge about the genesis of cancer, understand the disease as a multi-stage process. If not merely

[7] Another controversial debate, this one focused on the reference model itself, has been going for several decades. It developed in response to the research studies of A. Stewart (Green, 2003) and Gould and Johnson (Gould & Goldman, 1991), among others. But while the debate was fully developed and aired in international scientific journals, it was entirely ignored in French scientific circles.

external irradiation but also internal contamination at the level of a given organ is taken into account as relevant to the cancer process, it becomes necessary to attend to the effects of two types of exposure: extremely brief exposure of the entire body—termed acute exposure—and extremely low doses from point sources that hit very near target organs over a long period of time—chronic exposure—but vary in terms of internal movements and whether the organism responds by retaining or purging itself of inhaled or ingested radioactive particles.

Consequently, the point is to consider not only a succession of brief external radiation episodes—very poorly modeled by average dose over time, moreover—but also ongoing aggression of cells close to the place where radioactive dust has been retained, aggression that does not kill those cells but is sufficient to act at various stages in the genesis of cancer and cause various afflictions, namely at the level of the immune and endocrine systems.[8]

To support their hypotheses, the ECRR experts cite a considerable number of observation studies showing higher than expected numbers of cancer related to exposure to ionizing radiation. According to the committee, these studies reflect "dissonance" between the theoretical reference model used by the ICRP and empirical observation. This dissonance seems incompatible with the consistency that should obtain between a theoretical model and observed reality within a scientific paradigm. ECRR experts hypothesize that the pathological effects induced by low-dose radiation are more frequent and diversified than for heavy doses because of specific action mechanisms of the low-dose radiation when it hits various internal targets. One of the important differences between heavy and low dose effects is the fact that while heavy doses are lethal for cells, a dead cell is not a danger to the organism, in direct contrast to cells that have undergone mutation. Thus according to the ECRR, the likelihood of cancer developing in connection with ionizing radiation is much higher than that by the ICRP model.

The ECRR also criticizes the ICRP model for failing to take into account effects other than cancer, particularly those observed in populations exposed to Chernobyl fallout. They cite the general weakening of exposed persons' health, transgenerational effects (namely cardiac problems), hereditary genetic effects such as fetal and infant mortality, a set of non-specific effects and accelerated aging.

The ECRR is of the opinion that knowledge acquired over several decades invalidates the ICRP model and the radiological protection norms defined on the basis of that model, and they underline the urgent need to lower those

[8]As early as the 1970s, Rosalie Bertell (1985) called into question the mechanical reference model with its assumption that only an irradiated organ could set off radio-induced cancer. Bertell pointed out that irradiation could diminish the system's immunity and allow for cancer to develop in a non-irradiated organ containing a cancerous clone cell that had been produced by something other than radiation.

norms—by a factor of ten for annual radiation exposure among the general population (0.1 millisieverts [mSv] instead of 1 mSv) and by a factor of 4 for nuclear workers (5 mSv instead of 20 mSv).

Critical Discussion of the ECRR Recommendations at the French Institut de Radioprotection et de Sûreté Nucléaire

ECRR hypotheses calling into question the validity of the scientific reference model for health effects of radiation have elicited critical questioning at the *Institut Français de Radioprotection et de Sûreté Nucléaire* or IRSN, charged with defining radiation protection and nuclear safety rules in France.[9] A study group was organized at the Institute to debate the ECRR's hypotheses. It found that the ECRR hypotheses were not validated by animal experiments or "relevant" epidemiological studies, namely for internal radiation effects. But it specified that the absence of such studies is itself related to questions that have been "largely neglected by the scientific community" (IRSN, 2005). The study group came to the conclusion that in light of recent radiobiology studies it is necessary to revise the postulates on which the ICRP model was founded. The group acknowledged that there was a major error in how hypotheses on the effects of low-dose radiation had been reached: "It has long been postulated that incorporating 100 becquerels over a day was the equivalent of incorporating 1 becquerel over 100 days. This is perfectly accurate in mathematics, but false in biology" (IRSN, 2005, p. 21). Echoing ECRR hypotheses, the IRSN came out in favor of undertaking new studies on the possible existence of effects that are not taken into account in the model based on the cohort studies of Hiroshima and Nagasaki survivors.

While affirming that the ICRP reference model is still a reliable basis for the radiation protection norms in effect, IRSN representatives reached the conclusion that the scientific questions raised by the ECRR can no longer be eluded. In December 2005, in an interview in *Le Monde*, J. Repussard, IRSN Director, acknowledged the need to attend explicitly and systematically to uncertainties about the health effects of internal contamination. Commenting recently on the initial results of animal experiments undertaken at the Institute over the last four years,[10] Repussard notes that they show the chronic exposure system to be very different from the acute exposure system (*Le Figaro*, Mar. 17, 2006 and *Le Monde,* May 8, 2006).

[9]This public industrial and commercial institution was created in 2002. Its tasks are research and **expert advising**. It centralizes and manages all data on exposure to ionizing radiation as they pertain to radiation protection for workers and the population at large (www.irsn.org).

[10]These studies show that in the plants and animals tested, including rats, certain physiological functions—**in animals** the central nervous, respiration, digestion, and reproduction systems—are modified at low levels of internal irradiation.

Little mention is made in either the ECRR report or the IRSN commentary of exposure among nuclear industry workers, though they of all humans are the most likely to undergo daily exposure. The fact is that there is no consensus about the epidemiological data on the health consequences of chronic occupational exposure.

International Epidemiological Study of the Effects of Low-Dose Radiation Exposure in a Work Environment

From the first cohort study done in the United States, bearing on a population of 30,000 workers at the Hanford nuclear weapons production site (Mancuso et al., 1977, 1978, 1981; Morgan, 1980), to an international study recently conducted with 400,000 nuclear industry workers (Cardis et al., 2005), the relation between low doses of radiation and cancer mortality appears to have been scientifically established, though other studies continue to call that relation into doubt.

The cohort study of cancer-caused death among the 30,000 Hanford site workers, a study presented in the 1980 edition of the *Encyclopedia Britannica*, was for a long time the most important cohort study of an adult population chronically exposed to low-dose ionizing radiation (Morgan, 1980). In this study and those conducted in the next decades, the authors developed a scientific critique of the validity of the ICRP model of extrapolating effects of low doses from high-dose effects (Stewart & Kneale, 1982, 1993).

More recently, a retrospective mortality study was done, coordinated by the International Agency for Research on Cancer with state and nuclear industry funding support. The study covered civil nuclear workers in 15 countries, and its results, published in 2005 (Cardis et al., 2005), show a relative risk of mortality from all types of cancer except leukemia that is twice to three times higher among workers chronically exposed to low-dose radiation (19.4 mSv average combined dose) than expected on the basis of the linear no-threshold model derived from the Hiroshima and Nagasaki survivor cohort. The study authors are concerned to underline that this excess cancer mortality, while higher than that predicted by the model, is nevertheless statistically "compatible" with it.

In France these results have elicited opposite reactions from EDF management and the CGT union (*Confédération Générale des Travailleurs*) at EDF. The former cite the IRSN's affirmation that "these results do not invalidate the bases for the current radiation protection program" (IRSN, 2005a; *Comité d'Information des Professions de Santé,* 2005), whereas the CGT union has underlined the seriousness of the threat to workers (Grégoire, 2006). The union calls for reducing by a factor of three or four the individual exposure norm in effect, thus adopting the 2003 ECRR recommendations. It also points up the fact that the international study did not take into account employees of outside companies working in nuclear maintenance, who, as explained, take more than 80% of the collective dose received in French nuclear industry sites. It should be stressed that while collective dose has remained constant, the fact that it is being distributed over a greater number of exposed workers in the aim of reducing individual doses does not necessarily diminish the number of cancers to be expected from this exposure. The risk has simply been spread over a larger population. In any case, the division of labor that this book shows to be operative in the nuclear industry in France means that outside company employees bear the greater part of radio-induced cancer risk, whatever the official dose maximum.

In the initial feasibility study report for the international research project, *"contract workers"* were said to be a particularly important group for the study:

> To this day no study has been done in any country on cancer risk for these persons, because contract workers are a much more difficult population to identify and follow than permanent nuclear plant personnel. . . . Often because of the relatively high doses they receive in the space of a few days when operating in nuclear facilities, [these workers] are later assigned for a large part of the year to tasks that do not expose them to radiation. . . . Including them in the study of permanent personnel might introduce a risk estimation bias. (Cardis & Estève, 1990)

The point raised in the last sentence of this quotation refers to the hypothesis of some contract workers joining the permanent personnel: including them in the study might introduce bias since the only doses counted for them would be what they took as permanent personnel whereas in fact they would have undergone higher, longer combined exposure.

The definitive research plan of 1992 specifies:

> The definition of contract worker varies by country, as does the possibility of studying the cancer risk they incur. *Working under contract is thus a valid exclusion criterion.*[11] However, in all countries where this is possible, appropriate groups of contract workers will be included in the study population if there is a means of reconstituting their prior exposure and occupational past and of following them in connection with cancer risk. (Cardis & Estève, 1992)

Contract workers were therefore not taken into account in the international study, which for France bears on the permanent employee population, whom once again receive only 20% of the collective annual dose released in nuclear plants, whereas as this book shows, contingent workers are the most exposed to ionizing radiation and in France they are highly likely to "disappear" from company health surveillance programs. French rules stipulate that exposure of this population must be kept track of. It would have been entirely reasonable at the time the nuclear plants began to operate to set up a cohort survey of these workers, who perform tasks that are *habitually exposed* to radiation, by

[11] My italics.

collecting not only their external radiation dosimetry but also, for individuals, the radioactivity-exposed work activities they performed, instances and types of contamination undergone, and health problems appearing in the course of their occupational trajectories. If this had been done the international survey researchers would have had a much more reliable database.

The initial population for the international survey was 600,000 workers. In addition to "outside" workers, four categories were excluded even though they are part of the exposed worker population:

- 113,711 workers who had worked less than a year in a nuclear plant and for whom assigned tasks and the exposure conditions of those tasks were not specified;
- 39,730 internally irradiated workers, a type of exposure not taken into account in the ICRP model;
- 19,041 neutron-exposed workers: the pathogenic effects of neutron exposure may combine with those engendered by the λ radiation measured by classic dosimeters;
- 38,521 workers who did not wear dosimeters, though neither the reasons for this or the exposed activities they performed and that justified their being included in the cohort are specified.

Altogether, 32% of the initial population was excluded on the basis of criteria that give no reason to think that these workers were not exposed, not to mention the exclusion of high numbers of outside workers, the most exposed, as explained.

Other questions arise concerning the methodology of this epidemiological study and what it suggests about the workers excluded from the study population. The survey presented in this book provides access to accounts by "outside" company nuclear workers of their exposed activity, accounts that attest to the extreme variability of exposure profiles. This in turn renders dubious the validity of annual average dose as a variable likely to provide an exact image of exposure conditions, including for permanent nuclear industry employees. Lastly, the periods over which workers included in the cohort were exposed also vary by country and plant; some were exposed to ionizing radiation in their jobs as early as the 1940s, whereas others have only been exposed since the late 1980s. The latency periods between exposure and the appearance of cancer are such that it is reasonable to assume that in the most recently exposed groups, who may yet be being exposed, cancer is only just beginning to develop.

Leaving aside these critical remarks, it is important to note that the international study shows excessive cancer rates among workers who take a relatively low proportion of the collective radiation dose (for the French cohort, as mentioned, this is on the order of 20%). This points to a much more worrying situation for the "outside" workers who are the subject of this book. It must be kept in mind that results obtained for a population of stable, permanent workers performing production tasks cannot be extrapolated to outside workers performing maintenance tasks. The latter work intermittently; among them, external exposure to ionizing radiation is highly variable: they undergo greater exposure to radioactive dust and therefore greater internal contamination, multiple exposures to other occupational carcinogens and a combination of risks and constraints that has little in common with the working conditions of EDF permanent employees. There is a much higher probability of radio-induced cancers appearing in these workers. But it is also highly likely that this greater probability will not be observed if the only studies done are major statistical studies tending to make groups appear homogeneous that are in reality very different.

The fact that small, truly homogeneous high-risk groups are not followed, groups which each have a common memory that reflects members' qualitative knowledge of exposure conditions, is an obstacle to genuine observation of the real. And only such observation would allow us to provide the rigorously scientific bases for accurate comparison of the Hiroshima and Nagasaki cohort subjected to a single instance of extremely high exposure, and the various worker populations chronically exposed in extremely heterogeneous exposure conditions that need to be identified. The maintenance jobs presented in the early chapters of this book illustrate the heterogeneity of work performed under exposure to radiation.

2. MAINTENANCE, SUB-CONTRACTING, JOB INSTABILITY AND UNCERTAINTY, AND RADIATION PROTECTION: HOW ARE THEY EVOLVING?

There has been no revising of the decision to subcontract maintenance work in nuclear plants in France since publication of this book in 2000, despite the fact that the employment instability and health vulnerability engendered by this system are considered by many researchers two of the main determinants of worsened working conditions and, more radically, of the social construction of the invisibility of work-related harm to workers and populations at large.

The way that maintenance work and radiation protection organization in nuclear plants has been evolving has not affected the social phenomenon at the center of this book; namely the social division of labor and radiation dose and the "obliteration of traces" (see chapter 7) that makes it so difficult to bring to light the effects of this way of organizing labor on the health of workers and the general safety of nuclear facilities.

Increasingly Precarious Employment and Health: Invisibility and the Erosion of Rights

A number of research studies done in Europe, Canada, Australia, and Brazil underline the structural character of the transformations in work due to use of subcontracting and temporary agencies (European studies: BTS/SALTSA, 2000;

Seillan & Morvan, 2005; Canadian study: Lippel, 2004; Australian studies: Quinlan & Mayhew, 1999, 2000, 2001; Brazilian study: Franco & Druck, 1997, Druck, 1999). They also show the consequences of these changes for workers' health and how they diminish the effectiveness of arrangements for preventing occupational injuries and diseases from occurring and attaining compensation when they do. "Flexibility" has become the key word and strategy for organizing work in all countries, and it covers what is in fact a fragmenting of types of employment and activity that makes it much easier for large companies to reduce the number of their own employees to a minimum while retaining control over organization of production and maintenance. Subcontracting, outsourcing, limited-time contracts, temp work and "independent" labor designate some of the means of accomplishing this. Researchers have identified four key points of intersection between health in the workplace and what Quinlan and Mayhew call the "(dis)organization" of work.

These studies first point out the continual worsening over the last 20 years of existing unequalities in working conditions and the distribution of risk between permanent employees of dominant companies and outside and/or temporary workers with extremely diverse official statuses and subjected to more or less precarious employment. In a secondary analysis of the survey on living and working conditions conducted by the Dublin European Foundation, Daubas-Letourneux and I showed the inequality in exposure to work risks and constraints between European Union workers enjoying real margins of autonomy and EU workers operating within different types of work organization, characterized by flexibility and servitude and/or repetition (Daubas-Letourneux & Thébaud-Mony, 2003). Flexible work—the closest thing to what the Australian researchers call contingent work (Quinlan & Mayhew, 2000)—is far and away the type of work most likely to subject workers to combined health risks. According to the 2003 SUMER survey[12]/study, the most recent one on occupational hazards published in France, the activities in which workers are most likely to be at risk and to experience a combination of risks, notably exposure to carcinogens, are maintenance and upkeep, and the workers most likely to be at risk are those on short contracts, working for temp agencies, or on job training-placement programs (Guignon & Sandret, 2005). The first results of a survey being done with cancer patients in the Seine-Saint-Denis *département* in the outskirts of Paris show that most respondents have been essentially manual laborers, and that the youngest work in extremely precarious employment conditions and are very likely to be working in maintenance, upkeep and construction, where they are likely to be exposed to a combination of carcinogens (Réseau Scop93, 2005).

Second, the studies point out that the "flexibilizing" of work organization and the changes it entails are an obstacle to setting up legislative and regulatory

[12] SUMER: surveillance médicale des risques (medical follow-up of the occupational hazards).

texts and arrangements for preventing workplace-related health damage and obtaining compensation for damage. Whatever the country, researchers insist on the fact that these arrangements have not kept pace with the observed changes. They remain instead based on the model of an open-ended long-term work contract between employer and employee, and therefore do not cover any of the situations in which employees are obligated to work for several employers, only one of whom is their employer in the traditional work-contract sense. This means that often the labor rights that an employer is responsible for guaranteeing become inoperative. Quinlan in Australia (1999, 2001) and Lippell in Canada (2004) have shown the "erosion" of the right to have work accidents and occupational diseases recognized, an erosion linked to the way work is organized now.

Third, and as a consequence of the second point, all researchers underline the invisibility of work-related health damage due to the weakening of arrangements for monitoring and following up on the health of workers subjected to these forms of employment and for identifying what the work-related health damage is. Whatever the country, or system for recognizing work accidents and occupational diseases, it is extremely difficult to get compensation rights recognized and applied.[13] This recognition is even harder to obtain for workers in what are called "atypical" forms of employment because the conditions in which they are hired and work are not clearly covered by the rules governing recognition. Moreover, in statistical case-counting systems, accidents and occupational diseases are not officially listed in the category for the "principal" who prescribes the work, but rather in that of the victim's subcontracted employer. The institutional visibility of recognized cases is therefore distorted, because the party called upon to pay compensation for a work accident or disease is not the one actually responsible for the work activity in which the worker had the accident or was contaminated.

Fourth, the researchers insists on the radical threat to workers' right to union representation and freedom to speak about their working conditions. The gap is continually growing between the rights of protected workers who enjoy employment guarantees and forms of institutionalized representation and those of workers experiencing extremely various types of employment precariousness and whose relation to employment and work thus becomes individualized.

The reality studied in this book and developments in that reality fit into this research context. The increasing precariousness of employment caused by flexibility is more than ever a matter of social and labor tension in France, as

[13] In France a report by the *Cour des Comptes* [rough equivalent of the U.S. Government Accounting Office] blamed the agencies charged with recognizing work-related health damage for this situation and pointed to the low number of official occupational diseases in France (Cour des Comptes, 2002).

attested by the groundswell of protest in spring 2006 against a type of job contract particularly likely to make employment still more precarious. The generalization of work and employment flexibility—the "disorganizing" of work—works both to reinforce the social division of labor and risks and increase its combined effects on health.

What Changes In the Last 10 Years in the Situation of Workers Doing Radiation-Exposed Maintenance in Nuclear Plants?

Ten years have passed since the completion of our study and six years since publication of this book in France. In that time, the decision to subcontract maintenance has not been called into question. The main changes are a major decrease in collective dose received on nuclear plant sites and, for outside workers as a whole, a decrease in average individual dose levels to below the regulation limit of 20 mSv (IRSN, 2005b). Clearly this has been the result not only of European Union pressure on France to adopt the norms already in effect in the other EU countries but also of cooperation between researchers and labor union representatives to make visible the difficult, dangerous working conditions experienced by outside workers in the nuclear industry. Moreover, the fall in individual and collective doses must be considered in relation to how the overall context of maintenance work in nuclear plants has been evolving.

The study whose results are presented in this book brought to light three important dimensions of the kind of work organization in which outside workers get exposed to ionizing radiation. These are 1) managing employment and job assignment by radiation dose; 2) the invisibility of subcontracted work and of the associated occupational hazards; 3) the total lack of representation for outside workers on mixed management-and-labor union boards on occupational safety and health, called Comités d'Hygiène, de Sécurité et de Conditions de Travail or CHSCTs. Have these dimensions evolved since publication of this book?

Managing Employment and Job Assignment by Dose

In 2004, 31 174 outside DATR workers were on a list of workers to be monitored for exposure to ionizing radiation. The DATR workers had taken 2/3 of the annual collective dose received in French civil and military nuclear facilities, at individual levels that, according to the IRSN, did not exceed the regulation dose maximum of 20 mSv per year, except in the case of four individuals (IRSN, 2005b). How did the subcontracted companies handle the dose limit constraint in their organizing of maintenance operations? What happens to workers who have reached the regulation maximum exposure? The answer lies in the principle itself of managing employment and job assignment by dose; i.e., spreading dose over a sufficient number of workers to ensure that the limits are never exceeded.

The survey presented here shows the concrete modes used to apply this strategy, in which workers have to deal constantly with the contradiction between the constraints of getting the job done and the constraints of on-site radiation protection. Temporary and limited-time contract workers are most affected by this way of managing work, since their assignments or contracts are simply not renewed once they reach the limit. Are there any new data suggesting that employment is now less likely to be managed by dose?

In an epidemiological study of the working conditions and health of nuclear industry outside workers done by occupational doctors, those workers were interviewed twice at a five-year interval, in 1993 and 1998 (Doniol-Shaw et al., 2001).[14] In 1998, more than half the temporary and limited-time contract workers interviewed in 1993 could not be found. This confirms the facts analyzed in this book. Whereas outside workers are subjected to particularly sophisticated computerized surveillance, the exposure histories of those with the most precarious jobs—who are also probably those most heavily exposed to ionizing radiation—are not followed medically. And after a few years, it becomes impossible to find the workers.

We can hypothesize that the observed decrease in individual dose levels was obtained through drastic worker turnover at work stations directly and regularly exposed to ionizing radiation, namely through use of temporary contracts. Debates on dosimetry management at a conference organized by an EDF mixed management-union committee seem to confirm this (see Comités mixtes à la production, 2002). The 2005 report by EDF's Inspecteur Général de la Sûreté Nucléaire Pierre Wiroth also bolsters this hypothesis, underlining as it does that one preoccupying situation repeatedly encountered on nuclear sites during inspections is "workers not wearing a dosimeter" (IGSN, 2006, p. 22). As shown in Part I of this work, when the tension between contradictory demands is too strong, DATR workers are faced with choosing between leaving the dosimeter behind for dose-costly operations—meaning that the doses they receive during those operation remain invisible—and being forbidden access to plants after reaching the dose limit.

Invisibility of Subcontracted Work and Associated Occupational Hazards

The papers from the 2002 Comités Mixtes à la Production conference show the increasing concern of unionists at EDF about denial of real working conditions, including radiation exposure itself, and about the limitations of an exclusively quantitative approach to exposure. P. Davezies, a teacher-researcher in occupational health (Lyon 2 university, France) who attended the conference, explains: "Denying the specifically human dimensions of the activity can only

[14] This study is presented in chapter 6.

lead to forcing underground all worker mobilization aimed to offset the inevitable failures of technology. . . . If there is no discussion of means employed to handle a problem, the separation between prescribed work and work as it is actually effectively done becomes yet more pronounced" (Comités mixtes à la production, 2002, pp. 155-156).

Doniol-Shaw, an ergonomist and researcher at the Centre National de la Recherche Scientifique present at the conference, sees the particular characteristics of the work process in a nuclear plant as responsible for this separation:

> Can we speak of working in a nuclear plant without taking explicitly into account two characteristics specific to the work process on nuclear sites? On the one hand, the risk of nuclear accident and its consequences for the population and the environment, a risk that must be contained in the interests of general safety. On the other, the daily risk—not just potential but very real—of irradiation and contamination that goes with nuclear industry employees' activities and affects those employees' physical and mental health, a risk against which it is essential to protect oneself but from which one never totally escapes, especially if one is doing maintenance. (Comités mixtes à la production, 2002, p. 164)

Without citing this specific contradiction and the tension it causes for outside workers, the Inspecteur Général de la Sûreté Nucléaire at EDF, Pierre Wiroth, also underlines that in maintenance operations performed during scheduled plant shutdowns operators are "overworked" in order to get everything done that needs to be done in the limited time allotted. He observes that their working conditions "remain difficult overall," that "the quality of human relations between EDF personnel and service company employees remains a sensitive subject," and lastly that there is no "return on experience" that would allow EDF to profit from operators' analysis for better understanding "instances of malfunction" (IGSN, 2006, p. 39). These observations are in keeping with the results of ergonomics research done in the framework of confidential studies on work activity during nuclear unit shutdowns and are at times mentioned orally by these researchers at scientific meetings.

The accounts given by worker respondents in our survey give insight into different and often concealed aspects of how work exposed to radiation is organized. And the information on workers' situations obtained during the survey is hardly obsolete: more recent surveys and respondent accounts have confirmed it. It is further confirmed by the results of the forecited second STED survey (Doniol-Shaw et al., 2001). But the lack of medical and epidemiological follow-up have rendered lastingly invisible the combined effects of external and internal exposure to ionizing radiation as well as the effects of other work risks and constraints on the health of these workers charged with the daily upkeep and maintenance of nuclear facilities.

Developments in Rights to Representation for Outside Workers on Nuclear Sites

Created in France in 1982, the internal institution known as the *Comité d'Hygiène, Sécurité et Conditions de Travail* (CHSCT) gives active power to elected representatives of employees in private or public-sector companies or other work organizations in the matter of health protection for employees in their place of work. However, since the general switch to using subcontractors and temporary workers, there have been no changes to the *Code du Travail* to modify CHSCT composition and representation modes. When a major segment of work in the industry is subcontracted to outside companies that are required to have their own representative bodies, CHSCT prerogatives appear quite limited.

Union representatives at the *Comités Mixtes à la Production* conference made a strong demand for outside company maintenance workers to be represented on site-specific CHSCTs that would recognize the legitimate right of each category of worker to have their say in negotiations bearing on working conditions. The point was to recognize the fact that the activity of outside workers, together with the working conditions and risks inherent in that activity, are directly determined by the EDF principal in the framework of power relations between EDF and outside operator company employers. This demand is not new. In the 1990s a union organization created site-specific unions that were at the origin of joint labor actions by EDF personnel and site workers. But during the conference it became clear that in the absence of any formal legal obligation, there was little chance of such an institution coming into existence. The unions themselves did not take up this struggle, either in the nuclear industry or other sectors characterized by intense use of site-specific subcontracting.

A "Partnership charter" signed by EDF and its main service companies in 2004 created a consultative body on nuclear plant sites, the Commission Interentreprises sur la Sécurité et les Conditions de Travail [Cross-company committee on safety and working conditions] which provides for representation of outside company employees. But these committees are a place of discussion and dialogue; they have no real power. In contrast to CHSCT delegates, duly elected by their permanent-employee peers and protected by this status, the representatives of outside companies are not legally protected from dismissal. During the 2002 Comités Mixtes à la Production conference, a temp worker testified to being fired for having criticized nuclear plant working conditions. He indicated that other outside workers who had been involved like him in a labor action to protest against those conditions could no longer find work in the nuclear industry. This shows how narrow the margins for action are for outside company employees working in nuclear facilities, and it is consistent with what researchers specialized in the new employment precariousness have identified as the "erosion of rights."

On-Site Radiation Protection: Institutional and Regulatory Developments

Since the late 1990s, on-site radiation protection in France has changed in both institutional and regulatory terms.

Institutionally, all taskforces charged with defining radiation protection policy and monitoring protection in the framework of the general safety authority have been grouped together; radiation protection is now the responsibility of an autonomous institution that was created by inter-ministerial decision and is therefore free from any requirement of direct administrative allegiance to the central state and government. The IRSN is in charge of computerized dosimetry records for both permanent personnel and outside workers.

In regulatory matters, France has adopted the European Union on-site radiation protection norms and thoroughly revised government decrees on radiation protection for the population at large (*Code de la Santé Publique*, Article R.1333-43) and nuclear industry workers (Code du Travail R.231-81 to R.231-83).

These modifications have led to centralizing collection of information on exposure to ionizing radiation, now done outside the bounds of any social or political oversight. The ministries of health and labor have lost their specific prerogatives in the matter of radiation protection and the ministry of industry has acquired stronger influence in the fixing of public radiation protection policy.

The new provisions have not changed radiation protection for nuclear industry outside workers; it remains as presented and analyzed in this book.

CONCLUSION

The scientific (and political) debate on the effects of ionizing radiation is rooted in a struggle among experts, some in favor of and others against an epidemiological model developed more than half a century ago on the basis of an event—the dropping of American atomic bombs on Japan—that is unrelated to later forms of exposure to ionizing radiation; namely chronic exposure. The largest yet least visible population affected by exposure to radiation in the nuclear industry is the thousands and even hundreds of thousands of workers who perform daily maintenance and upkeep operations and tasks in nuclear plants, nuclear testing facilities, research centers, reprocessing plants and nuclear waste management centers. The dismantling of nuclear reactors is already another important cause of exposure to ionizing radiation, this time among workers doing the job of demolishing radioactive facilities, and it will continue to be so long into the future. Outside the nuclear industry, industrial or medical applications of radioactivity have greatly increased occupational exposure to ionizing radiation, and many of the exposed workers in this case are virtually unaware of the health risks they incur.

France's nuclear facilities are half a century old, and with the exception of dosimetry figures, no specific arrangements have been devised to record the "memory" of work and internal and external exposure to ionizing radiation in the French nuclear industry. The dangers of radioactivity, like those of asbestos, have been known for decades**; in that time** the "traces of the dead" have begun to be transformed into the "death of all traces," to use Louis-Vincent Thomas' expression, quoted in the epigraph to this preface. The same conclusion was reached by the journalist J. P. Desbordes in his *Atomic Park: A la recherche des victimes du nucléaire* [Atomic Park: in search of the victims of nuclear power] (Desbordes, 2006).

This book is intended above all as a contribution to the work of memory that is or should be the foundation of our knowledge and understanding of industrial risk. The work of memory involves collecting—and listening to—what those most directly affected by such risk and the meaning of working in this high-risk technological industry have to say. As explained at the end of the book:

> The maintenance function is the indissociable "underside" of nuclear reactor operation. Reactor operators cannot practice their skill unless that skill is "secured" (as we speak of being "secured" in mountain-climbing) by the skill of the "outside" workers who maintain nuclear installations, repair them when necessary and labor to prevent breakdowns. Maintenance work too draws its meaning from the contribution made by "outside" workers to this "life and death-bearing" activity, for it is an activity that all nuclear industry workers assume responsibility for. These workers commit their physical persons. They take on the unavoidable danger inherent in their activity—the danger of losing health and life—so that others—all others, EDF permanent personnel and society at large—may be protected, at least for a time, from deadly exposure to ionizing radiation.

Introduction

This book lies at the intersection of two issues that are of crucial importance for public health: the threat posed to health and safety in the workplace by the practice of subcontracting, and the management of major technological risks.

In the nuclear industry more than in any other, a particular risk is of course dominant, omnipresent: the risk of atomic radiation, with its deleterious effects on health. And more than in any other industry, to monitor this risk, the choice has been to use technology and automatic systems. The choice points to a fundamental debate about industrial organization. Technicians and risk management specialists claim that technology itself is what makes industrial equipment and installations safe. This also applies to surveillance of what in their view is the weak link in the technological process chain: human intervention. Analysis of work activity by researchers in the social sciences of work, however, shows that the production process continues to be controlled by operators and groups of workers whose activity is crucial to making machines and automation work. Guy Jobert's anthropological approach to job performance recognition for nuclear power plant control operators (1999) and Yves Clot's psychological study of work places and living places (1995) brought to light the changes in work procedures that have accompanied computer and automation-assisted production. Operator activity now consists less in producing than in making machines produce. But this kind of work, this version of "efficiency despite all," to use Clot's expression, tends to remain invisible as long as the machine is functioning normally. It is only when the machine breaks down or an accident occurs that the indispensable role of human labor becomes apparent.

The complexity of modern industrial socio-technical systems such as that of the nuclear industry masks the work of the producers themselves. It tends to conceal the contradictions that producers have to deal with in order to make production possible—possible without risk to themselves or the populations that would be adversely affected by an accident. The point of the nuclear industry is to produce electricity. Who are the "producers" in this industry? That is, who are the people who make possible the "efficiency despite all" of the nuclear production process? Are they the engineers who develop and analyze operating modes? The managers who preside over the organization of human labor? The control operators who oversee regulation of the instruments that pilot the functioning of the nuclear reactor? Or the "outside" workers employed by subcontractor companies who, in the shadows, on the "underside" of the system, ensure maintenance of equipment and installations, prevent wear and tear, and do what's necessary to avert breakdowns and accidents?

The problem studied in this book lies at the intersection of my own research studies and studies in different areas of social sciences that in turn clarify the approach used here.

THE "UNDERSIDE"? "OUTSIDE" WORKERS?

To understand the problem, it is essential to look at the meaning of certain words. Why call what is at the very core of nuclear production and safety issues—namely, the maintenance of nuclear plant installations—the "underside"? Why use the term "outside" to designate workers who participate fully in the process of nuclear production, who make that process possible by regular interventions that are absolutely necessary to keeping installations in working order? These two questions are closely linked to the one that has oriented the research presented in this book: In the nuclear industry—in France but also outside France[1]—why is installation maintenance ensured by what are called "outside" workers, some of whom are under temporary contract or working for temporary agencies?

In a study I conducted in the early 1980s of workplace accidents among "outside" workers participating in maintenance operations in another cutting-edge industry—electronics—I brought to light the social division of labor and risks engendered by subcontracting (Thébaud-Mony, 1984). In that study, long-observed criteria of differentiation—the fact that the persons most affected are manual workers, often of foreign origin—were found to go together with a profound split between permanently employed operators working in production and subcontracted workers, called "outsiders," in charge of maintaining, cleaning and guarding installations. Accounts collected from "outside" workers on this industrial site testified to a distribution of risk along clear social and labor lines: on the "topside;" i.e., production, the image is of perfect control of the technical process and associated risks, whereas on the "underside" we clearly see the limits of this control, as revealed by workplace accidents.

[1] There are very few studies of the organization of maintenance work in the nuclear industry. M. Bourrier's doctoral thesis in sociology of organizations alludes to the use of temporary workers in American nuclear plants. Otherwise, a European Union directive aimed at ensuring "the operational protection of 'outside' workers exposed to ionizing radiation when intervening in the controlled zone" was adopted in 1990 (see p. 155 for a discussion of this text)—evidence that in various European Union countries the problem was considered a structural one. Lastly, a study of subcontracting in the Japanese nuclear industry (Tanaka, 1997) brings to light a situation in that country that is very similar to the one in France.

When these results were brought to bear in a comparative study of trends in workplace health and safety in France and Brazil, it became clear that this is a structural phenomenon present in different forms in these two countries (Thébaud-Mony, 1990).

The very term "outside" worker appears in official regulation language. The French government decree of November 1977 (revised in 1992) gives "workers safety prescriptions for intervention by 'outside' companies working on the site of a contractor"—as if what were at issue was a new category of business company and employee. In the meantime, subcontracted work and the accidents associated with it are made invisible.

Socio-Economic Insecurity and Subcontracting

In the framework of a "thematic scientific action" program at the Institut de Recherche sur les Sociétés Contemporaines (IRESCO) organized in 1995-1996 on the initiative of research teams at the Centre National de la Recherche Scientifique (CNRS) and the Institut National d'Etude sur la Santé et la Recherche Médicale (INSERM), a multidisciplinary approach to the theme of "increasing insecurity in employment, work, and health" was applied to study of relations between changes in production organization and increased insecurity in hiring, work, and health in the workplace. Central to this analysis, which called into question the classic understanding that business modernization has marked the end of the Taylorist division of labor (Appay & Thébaud-Mony, 1997), was the transformation of production organization and social relations. Modernization thus defined, presented by sociologists and economists as an alternative to Fordism, was founded on the Japanese "model" of just-in-time production, where the favored and indeed crucial instrument for modernizing companies is flexibility (Kern & Schumann, 1989; Piore & Sabel, 1984). All types of specialists of work—sociologists, economists, ergonomists, and psychologists—questioned this postulate on the basis of what they observed of working conditions, particularly the concealed aspects of the Japanese "model"; namely, subcontracting and the use of temporary workers (Hirata, 1992). In France in the 1970s and 1980s, flexibility became a major tool for restructuring businesses. The abstract term "flexibility" masks the real material phenomena it refers to: intensified work, break-up of worker collectives, lay-offs, structural unemployment, use of temporary workers and subcontracting. Temporary work was legalized in France in 1972; subcontracting in 1975. Both these policies ran counter to the century-long effort to construct labor law on the basis of work contracts—a century during which jobs' trade was strictly prohibited ("Subcontracting" file, 1993; Thébaud-Mony, 1994). As early as 1978, a study published by the Ministry of Justice underscored the threat that subcontracting poses to the guarantees associated with work contracts: the employer company is in a position to "freely modify the framework within which mutual rights and obligations are exercised" (Rettenbach, 1978). Indeed, for employees of a subcontractor company that is acting for a contractor, labor regulation comes under the sway of what can only be called the market relation between user-contractor and subcontractor companies. Fifteen years later, Alain Suppiot underscored the now confirmed existence of "a fracture line running through labor law between workers who benefit fully from the personal rights guaranteed by a 'typical' work contract and those who become more like 'labor merchandise' due to atypical work contracts" (Suppiot, 1994). In his chronicle of employee status, Robert Castel (1995) traced the historical and social construction of the "deterioration of wage and work conditions." His point was that by pushing out their most senior employees and making it difficult for the young to enter, companies were working against their "integrative function"; they were "dualizing the labor market," and "destabilizing stable employees." These are the social consequences of flexibility when it becomes the dominant value of the production system.

In political discourse and scientific research into socio-economic and employment-related insecurity, however, flexibility generally appears as the "social scourge" of modern times, affecting groups considered vulnerable. Following this understanding, research came to focus on what characterizes groups that fall victim to socio-economic insecurity and exclusion (Haut Comité à la Santé Publique, 1998; Paugam, 1996).

What gets lost in these approaches to the phenomena of flexibility and socio-economic insecurity is the indissoluble link between the two, namely the social organization of labor. At the interface between "modernization" of production systems as induced by flexibility and the structural insecurity that a growing proportion of industrial societies are now locked into, what we have is an ongoing process in which people are made socio-economically insecure—a process which for more than 20 years has been increasing the degree to which working men and women are subjected to companies' needs to be productive and competitive. This is shown in a set of studies in two issues of *Actes de la Recherche en Sciences Sociales* devoted to "new forms of domination in labor" (1996), and in various articles in a collective work entitled *Précarisation sociale, Travail et Santé* (Appay & Thébaud-Mony, 1997), as well as in Beaud and Pialloux'study of the automobile industry (1999).

Subcontracting and Workplace Health and Safety

The best-known study of the use of subcontracting as a tool for externalizing risk is by the German journalist Günter Wallraff (1985). It showed the social invisibility of "temp" and subcontracted workers, precisely those who handle the most dangerous tasks in German industry.

In France, several studies by the Institut National de Recherche et de Sécurité (INRS) brought to light the disproportionate number of accidents associated with outside and temporary workers (François, 1991; Vandevyver, 1984). These results were confirmed by a national survey conducted by the French Ministry of Labor's Observatoire des Risques Professionnels (Cristofari, 1994). In the conclusion of a report by the Conseil d'Etat on work accidents in the building industry, Max Querrien calls for prohibiting sub-subcontracting as a means of market regulation, explaining that this is the only truly effective way of improving working conditions and diminishing the frequency and seriousness of work accidents in this sector (Querrien, 1991). These recommendations have not been followed. Two other studies by the INRS on differential exposure to toxic products, one on exposure to metallic catalyst dust in the chemical industry (Hery et al., 1994), the other on on-site exposure to chemicals (Hery et al., 1995), brought to light the phenomenon of externalizing risk onto outside company personnel charged with industrial maintenance and cleaning. However, there is no national statistics source to provide systematic access to information on workplace accidents and occupational disease among outside workers employed by subcontractor companies. Information on work accidents is available in contractor records, but it is not taken into account by the Caisse Nationale de l'Assurance Maladie [CNAM; French national health insurance bureau]. In the annual workplace accident and occupational disease statistics compiled by the CNAM, no mention is made that would allow for identifying work accidents involving subcontractor or temporary agency employees intervening on contractor sites. But the existence of a relation between subcontracting and work accidents was well enough known and acknowledged in 1992 that Labor Minister Martine Aubry, speaking at the 3rd "International Forum on Health and Safety at Work," denounced the former practice as unacceptable "subcontracting of risk."

The division of labor and the risks associated with subcontracting represent a structural dimension of industrial production organization characteristic of the 1980s and 1990s, as attested by certain international publications. In Sweden, despite certain methodological difficulties, the phenomenon has been illustrated for the mining sector on the basis of national work accident statistics (Blank et al., 1995). In Finland, a study of the influence of organizational factors on occurrence of serious accidents brought to light inequality in work accident risk between subcontracted and permanent contractor workers. The accidents were linked in particular to extremely tight time constraints and schedules and serious safety hazards (Salminen et al., 1993).

Risk distribution is not the only effect of subcontracting. Subcontracting also modifies social relations in the workplace, as well as the conditions in which safety rules and regulations are applied. This is shown by an Australian study of four activity sectors on the effects of subcontracting on worksite health and safety (Mayhew, Quinlan, & Bennet, 1996). In addition to a higher proportion of accidents among subcontracted workers, the authors bring to light three levels of effects that subcontracting has on worksite safety and health conditions. First, given the competition among subcontractor companies—competition that is of course fueled by contractor companics—subcontracted employees are required to perform under strong time constraints and to justify their work to a variety of hierarchical superiors. This greatly intensifies their work. Moreover, the complexity and intertwining of levels of accountability and safety surveillance tend to make a muddle of safety rules and constraints and thus impede implementation or application of them. Lastly, geographic mobility and time flexibility greatly worsen conditions of family and personal life for subcontracted employees.

Two studies, one French, the other Japanese, take up the specific question of on-site subcontracting. The French study observed different chemical and nuclear industry sites and was conducted in collaboration with researchers and labor union representatives. The authors distinguish four major categories of "organizational risk" having a direct negative influence on "outside" workers' health: constraints built in to business contracts between contractor and subcontractor companies, since subcontract company workers are in no position to negotiate those constraints; the externality of the subcontracted company and its employees in relation to the contractor, which impedes risk identification and control by those actually doing the work; the pressure of time constraints, which affects both conditions in which the work is done and employees' personal work paces; lastly, "nomadism" among sites, which causes physical fatigue and psychological fragility, since employees have no regular family life (Pereira et al., 1999).

"Nomadism" is particularly emphasized in the Japanese study, focused specifically on subcontracting of maintenance tasks in the Japanese nuclear industry (Tanaka, 1997). As in France, the tasks of maintenance, decontamination, and repair work in irradiated zones of nuclear plants fall to the "gypsies" of the nuclear industry (the author's term). The characteristics of maintenance work organization in Japanese nuclear plants are very similar to those in French ones: sub-subcontracting; marked inequality between permanent and subcontracted workers when it comes to exposure to radiation and radioactive contamination; employment threatened if regulation-specified radiation dose is exceeded; intense mobility of workers among the various nuclear power plants. There does therefore seem to be a specific approach to nuclear risk management, one in which subcontracting plays a major role.

INDUSTRIAL RISK MANAGEMENT AND SOCIAL RELATIONS IN THE WORKPLACE

Perceptions and management of the health risks associated with high-risk technology such as nuclear energy production are shaped by several systems for interpreting what, in work organization itself, can constitute accident or incident dangers.

"The Alert"

The CNRS program entitled "Risques collectifs et situations de crises" has made it possible to research social representations and perceptions of risk. These studies were presented and discussed in the framework of a seminar led by Claude Gilbert (see *Actes du Seminaire du Programme "Risques collectives et situations de crises,"* 1998). One important study is Francis Chateauraynaud and Didier Thorny's analysis of processes of "alert" initiated by people whom the authors call "grim precurseurs." This "alert" procedure has caused actors, skills and major scientific controversies related to risks and risk-provoked crises to emerge into the public space (Chateauraynaud & Thorny, 1999). For health and safety in the nuclear industry, these controversies are of major importance, particularly the one concerning effects of low doses of ionizing radiation because the question of whether or not a low-dose toxicity threshold exists directly influences protection of both nuclear workers and the surrounding population. The authors do not directly examine work activity itself, though they did do a case study of some "alerts" as related to the use of temporary workers in the nuclear industry.

Human Error

Major technological risk management raises questions about accidents and their causes, since operators in the field are the first actors to be suspected of human error (Laville, 1991). In the "technicist" approach, which follows a typology established by Nicolas Dodier, research into the causes of accidents and incidents focuses on human error resulting from operator behavior (Dodier, 1998). This approach to labor psychology is based on the postulate that the technical and technocratic experts who define safety rules and procedures are infallible. The idea is that the scientifically sound theoretical models which provide the foundation for such rules and procedures cannot be wrong or misguided because the professionals who conceived them are designers of nuclear installations and experts in the recent discipline called risk management science. Under the egis of the French Academie des Sciences a colloquium was held in 1998 in Paris on the theme of "risk and society." It forefronted specialists from the Institut Européen de la Cyndinique [cyndinics: the science of risk], a discipline that "for the last ten years has been working to formalize general principles for accident situations in the aim of developing rules for preventing and managing risk. . . . Danger sciences have borrowed their models in large measure from the general safety culture developed in the nuclear industry."[2] In cyndinics, prescriptions and prescribers are understood to be infallible because their approach is based on scientific and technical knowledge of how installations function. In this view, any and all incidents can only be due to operators' "human

[2] *Le Monde,* Dec. 4, 1998.

error." The weak link in the nuclear industry, the safety enemy, then, is operator fallibility, which explains the absolute necessity of "training" operators to act responsibly and inculcating a safety culture. This is a behaviorist ideology that defines workers—here essentially "outside" workers—as "high-risk" individuals and understands incidents or accidents as due exclusively to individual behaviors that do not comply with prescribed norms, rules, and procedures.

James Reason (1990) called this theory into question, providing a cognitive psychology view of the hidden springs of human error. Drawing lessons from past accidents, particularly in the nuclear industry, he brings to light two types of "universal human failings" which he designates "fundamental attribution error" and "fundamental surprise error" (p. 212). The first "refers to a pervasive tendency to blame bad outcomes on an actor's personal inadequacies . . . rather than attribute them to situational factors beyond his or her control" (p. 212). The second "is to avoid looking at any fundamental meaning and to learn the situational lessons from the surface events" (Lanir, 1986, p. 213). Reason cites Woods (1987), "Prior to TMI the industry could and did think of nuclear power as a purely technical system where all the problems were in the form of some technical area or areas and the solutions to these problems lay in those engineering disciplines. TMI graphically revealed the inadequacy of that view because the failures were in the socio-technical system and not due to pure technical nor pure human factors" (p. 214).

The behaviorist thesis thus has severe limitations, which may be said to lie in the blind spots of an exclusively psychological approach to relations between humans and technology.

Safety as a Social Construction

In sociology of organizations, the main interest has been in the dynamic character of the organizational processes within which reliability of high-risk industrial systems is constructed—or destroyed. This is the main focus of a book by Mathilde Bourrier on maintenance activity organization in four nuclear plants in France and the United States (Bourrier, 1999). Her work makes use of two theoretical approaches, Charles Perrow's global institutional perspective (1984) and the "tension in work activity between autonomy and prescription" which is the focus of numerous studies in ergonomics and sociology of work. For Perrow, industrial accidents are the "normal" (unavoidable) product of the types of institutional logic operative in the organization and functioning of complex systems.

Referring to the main hypothesis of Gérard de Terssac (1992) on the place of human beings in automated functioning, a place that they owe to their ability to "produce the informal rules and organization solutions absolutely required by production," Bourrier points out the limitations of this idea, hypothesizing that organizational arrangement itself constitutes a decisive factor in what strategies operators devise and use. She forges the concept of "organizational reliability," which allows her to base evaluation of the functioning modes of the organizations she studies on examination of both individual and collective practices and organization strategies. Specifically, she works to "understand how individual and collective actors work every day to accomplish their tasks, negotiating their participation in a highly constraining form of work organization." Her study of how maintenance is organized and carried out in the four nuclear plants yields a rich analysis of how these two components are interrelated.

The comparative approach also enables her to identify important differences between French and American plants with regard to reliability of maintenance organization and therefore safety, particularly through her observation of power relations and their influence on operator activity. Whereas in American plants the power to modify rules and prescribed procedures is "shared" between managers and hands-on operators, in France "no arrangements have been devised to help task performers modify [rules or prescribed procedures] or get those rules modified when they run into problems." Given the French practice of sub-subcontracting maintenance tasks,[3] the impermeability between prescription and activity creates total opacity around the real work that maintenance workers do.

Bourrier concludes that "the maintenance act is a socially constructed activity," explaining: "The act of following or not following procedures which seem to commit only one individual or at most two or three engaged in a particular task, is in fact not an isolated act performed by a person caught up in his or her own psychological motivations. Hidden behind this solitary act is the entire thicket of social relations characterizing the organization. The possibility of following or not following a procedure is therefore in large measure socially constructed; that is, produced by the organization." She also pleads for a critical analysis of this "thicket" and of the concrete conditions in which maintenance is carried out, as the only means of evaluating the real conditions of reliability and safety in nuclear plant functioning.

Industrial Accidents as a Sociology Research Problem

On the basis of an international comparative study of a century of work accidents in industry, Thomas Dwyer (1991) developed a critical sociological theory of technical and technocratic conceptions of security, safety, and accidents.

[3]It seems common practice in American plants to use temporary workers, but according to Bourrier's information, this involves direct contracts between plants and workers in one of three forms: independent workers (in particular automation specialists and "jumpers" [nozzle dam installers]); professional skilled worker unions (unions are in charge of bringing together supply and demand, but there is still a direct contract between the plant and the union worker); and "job shops" which serve as supply and demand intermediaries and ensure management of work contracts (similarly to French temporary agencies).

A conflict of legitimacy exists between the worker, who is concerned to ensure his right to a job without risk, and the employer, who thinks in terms of the economic "health" of his company. In the late nineteenth century, with the passage of workplace accident laws, the justice principle yielded to an approach to accidents in terms of insurance and technology. This gave the technical experts the power to say what was what in prevention of occupational risk and accident reparation, and it was in this way that "social peace" was constructed, in the form of a compromise between management and the unions on the matter of health and safety in the workplace. Nonetheless, beneath this peace the legitimacy conflict persisted, identical to what it had been in the past: the conflict between the right to health and safety in the workplace and economic rationality.[4]

For Dwyer, industrial accidents are "produced" by social relations in the workplace at three levels: "rewards," which range from wages to signs of symbolic recognition and esteem and include working hour arrangements and working conditions; "command," that is, the forms of power that management exercises over workers in matters of work management; the "organizational level," which involves forms of organization and labor division as well as management-imposed task distribution. Dwyer calls the main factor in industrial accidents "disorganization."

There is also the specific dimension of the individual, his or her personality and history. Dwyer also questions the psychological approach to interpreting accidents which holds that the causes are to be looked for in individuals' characteristics and personality traits. In his view, this approach lacks two fundamental dimensions. First, individual histories are rooted in social history: you cannot isolate a worker's "personality" from what fashioned that personality. Second, the various levels of social relations identified above, and the related forms of domination, determine at all times the maneuvering room and margin of legitimacy that circumscribe individual practices, behaviors, and attitudes. More structural relations (concerning skill, sexual division of labor, subordination in terms of employment, personnel status, or wages) work to determine the situations individuals experience and whether individuals and groups do or do not have any maneuvering room in their work.

Throughout his book, Dwyer relates the history of health and safety in the workplace to that of the trade union movement. These histories are also present as a kind of backdrop for any attempt to understand social processes of risk management in the workplace. "Social peace" between management and the union movement around questions of health and safety was constructed through the triumph of the technical approach to accidents and accident prevention. The unions were (are) prisoners to the myth of the extraordinary power of technical

[4]On this point the results of my studies of health in the workplace and occupational disease in France are convergent with Dwyer's conclusions (see Thébaud-Mony, 1990, 1991, 1994b, 1997a).

progress, particularly its ability to make most industrial accidents disappear—all by itself, independently of social relations and the kinds of political choices generally associated with technology use, the idea being that whatever risk of accident remains can be corrected by appropriate worker training. Correlatively, and consistent with the precise logic of the technical progress paradigm, the unions were also dominated by the ideology of "normalization," which "frees" general and workers safety issues from a social approach to risk and submits them instead to a technocratic approach holding that consensus between labor partners is to be established around norms and danger thresholds through diagnosis by experts invested with scientific legitimacy. This approach conceals implicit value systems and the choices they produce in determining what is called "socially acceptable" risk.

But Dwyer also observes the "rupture in social peace" manifested when public health disasters such as nuclear accidents occur. He locates the fracture line not between management and employees (i.e., between capital and labor) but between the employees body itself and the unions. Union organizations occupy positions that make them party to the social construction of compromises for adopting "socially acceptable" risk levels. However, most of the people directly exposed to risk and working under specific conditions and constraints, those who suffer the consequences of that danger, have no recourse—not even union representation—for expressing themselves publicly. For Dwyer, the social construction of workplace accident invisibility develops in this "rupture" and in the underlying social relations of domination, a system of domination that is bolstered by the ideology of scientific, technical, and technocratic management of occupational health.

This approach to exploring the ties between subcontracting and occupational health in a high-technology industry such as nuclear power requires encompassing all dimensions of the social organization of labor, together with the ideologies and social relations that determine that organization, in the research process. The researcher, meanwhile, must stay as close as possible to "outside" workers' lived experience of their activity.

RESEARCH METHODOLOGY AND PROCEDURE

The aims of the present research study are:

- to understand why an industry whose economic credibility is founded on infallible technical monitoring and control of radioactive risk subcontracts one of the most essential monitoring and control functions, namely nuclear installation and equipment maintenance;
- to analyze the consequences of this on the health of "outside" workers and the general safety of nuclear installations (the effects, therefore, on the health of the neighboring population). Given the weight of the nuclear industry in France's energy policy today, these are major public health issues.

It should be underlined that more than 80% of maintenance operations are performed by more than a thousand subcontracting industries intervening in France's basic nuclear installations. Every year, 20,000 to 30,000 outside workers are "directly assigned to tasks that expose them to ionizing radiation" [Directement Affecté aux Travaux sous Rayonnements; henceforth DATR] in these plants and undergo special medical surveillance for this risk.

Preliminary Stage

At the request of the Committee of occupational safety and health (*Comité d'Hygiène, Sécurité, Conditions de Travail*, henceforth CHS-CT) of the Chinon nuclear power plant, a survey was conducted in 1989-1991 of regulation medical surveillance of the dosimetry of employees performing services on basic nuclear facilities (Thébaud-Mony et al., 1992).[5] The results show a massive transfer of radiation exposure risk from contractor employees onto subcontracted workers: outside company workers received 80% of the collective annual radiation dose received in France's nuclear plants,[6] and at high radiation exposure levels: on average, outside employees are exposed to 8 to 15 times more radiation per month than EDF's DATR employees. The study also brought to light extreme individualization of occupational itineraries (this is very hard to reconstruct given the common practice of sub-subcontracting, job instability and insecurity, and worker mobility). Lastly, the fact that workers were given medical examinations at so many different sites itself attests to the difficulty of really following them;[7] medical examinations of these employees became a sort of "ritual" for discerning fitness for the work in question, disconnected from ongoing health surveillance and protection.

An epidemiological survey of "outside" DATR workers by worksite doctors ("Sous-traitance EDF-DATR or STED)" has allowed for identifying statistically verified constants in these employees' working and living conditions (Doniol-Shaw et al., 1995). In addition to the high levels of radiation received by temporary workers, particularly in certain jobs, the authors underscore the strict time and work-pace constraints weighing on these workers, as well as the immense physical effort demanded of them. A statistically significant relation was found between exposure to these constraints and the health problems cited by employees

[5]This refers to recording of radiation doses received by operators working in controlled zones of the plant; i.e., exposed to radioactivity.

[6]This percentage has remained constant since then and is confirmed by official Electricité de France (EDF) and Office de Protection contre les Rayonnements Ionisants (OPRI) figures.

[7]The 87 employees we surveyed over 4 years were followed by no fewer than 103 doctors. Some of them never saw the same doctor twice for medical follow-up.

(digestive ailments, spinal cord problems, sleeping difficulties, mental, and nervous exhaustion). The authors insist on the decisive influence of work organization factors.

The results of the two studies are consistent with each other, particularly on relations between job insecurity and the different proportions of radiation received by the different types of employees—the phenomenon that researchers have called "managing employment by dose." This type of work organization, presented in detail further on, diffuses the risk of radiation exposure among a high number of workers scattered throughout France. How can long-term medical tabs be kept on persons exposed to a danger that may have such deferred effects as leukemia or cancer? Moreover, radiation risk is part of the closely interrelated set of constraints associated with both specific occupational pathologies and signs of premature aging.

The two studies cited do not answer the question about the legitimacy of and reasons for subcontracting maintenance work in the nuclear industry and do not comment on what they reveal of the extremely trying living and working conditions for outside workers, incomparable to those that may be observed when EDF personnel, i.e., permanent contractor company employees, are surveyed (though there are disparities within the EDF group). However, hypotheses derived from these first quantitative results have allowed me to conduct a qualitative sociological study not only of the consequences of this way of organizing maintenance work but the very principles of its construction and legitimacy, among them the role played by health-and-safety issues.

Hypotheses

At the center of this research process are the following two hypotheses:

- Subcontracting risk and "managing employment by radiation dose" work to conceal the social contradictions inherent in the production of nuclear energy;
- The choices involved in the social organization of maintenance work in nuclear power plants have heavy consequences, either immediate or deferred, for "outside" workers in this industry, for the safety of nuclear plants, and therefore for the health of the surrounding populations at large, who are at risk for radiation and environmental contamination not only by a possible nuclear accident—obviously this is the most dreaded danger—but also and more immediately by various (hidden) forms of radioactivity dissemination.

Research Methods

On the one hand, I needed to accede to what was hidden—the social contradictions—and to the real work of nuclear plant maintenance. This was done by collecting accounts from "outside" workers of the way their hiring, working conditions and activities are organized. The book considers the work both as it is prescribed and as it is actually done, the latter accessible once again through "outside" worker accounts. This approach was particularly necessary given that these employees are indeed "outside workers": the legitimacy of their presence in the nuclear plant can be called into question at any moment. Ergonomics specialists, philosophers, and linguists agree on the difficulty of speaking about work (see Boutet, 1995), and the accounts collected here are not exempt from this difficulty. They nonetheless have provided access to the lived experience of nuclear power plant maintenance activity, the social contradictions with which these "outside" workers are faced, and their perceptions of their work. In particular they have made it possible to understand what it means to "work under exposure to ionizing radiation."

On the other hand, once the social organization of the work was apprehended, it was necessary to determine the foundations of its social legitimacy within industrial institutions as well as in the state. The state is heavily implicated in both the evolution of occupational health regulations, particularly radiation protection for workers, and of course public policies for ensuring that workers health and safety regulations are applied. It also plays a major role in research choices and decisions.

Lastly, once it has been shown how subcontracting maintenance enables the nuclear industry to sidestep its own contradictions, it is important to examine the consequences of this practice on public health (health of industry workers, surrounding populations) and possible means of assessing those consequences.

The sociological approach to occupational health that underlies this analysis of workers' relation to the risks of radiation and contamination implies a wider vision of the problem than just the individual's exposure to ionizing radiation. We are considering workers exposed to a risk for which a full set of precautions has been defined; we therefore necessarily go beyond individual experience of this danger, seeking to understand how it fits into the dynamic of work-health-employment relations and what latitude employees have for protecting themselves while coping with their other work constraints. The point then is not only to identify the risks and constraints a person experiences, but also, by means of a wider-reaching approach, to reconstitute the social processes that constitute the individual's history and construct his or her health and future. This dynamic involves theoretical elaboration of a definition of health that differs from exclusively biological fundamentals and puts less emphasis on the notion of the state (though these remain the dominant references in public health). Breaking with a conception of health in terms of "absence of disease," the World Health Organization defines health in terms of an ideal "state of complete physical, mental, and social well being and not merely the absence of disease or infirmity."

Using a dynamic, relativist approach, Georges Canguilhem opened up another view in his renowned text, *Le normal et le pathologique*, defining health as a process. Philippe Davezies uses this dynamic view to determine the different types of effects that working conditions and types of work organization may have on health: "Damage to health in the workplace may be due to the direct action of harmful physical or specifically chemical materials, to inappropriate use of men and women in the workplace, or to damage to sense of dignity and self-esteem." These different ways in which health is endangered or harmed in the workplace may be observed by means of differentiated methodological tools, and this leads in turn to various strategies for health-related action at work, strategies focused not only on material working conditions but all dimensions of work organization (Davezies, 1998). This approach to occupational health opens onto the sociological perspective operative here, where health is considered a social construction in which social roles play a determining role. Health as defined in this work, then, is a dynamic process in individual life, a process that imprints the effects of an individual's work, living conditions, pain, pleasure and suffering on his or her body. It is the trace of all the individual's history is made of, though the individual history is also a collective one because of the many social influences that shape it.

Several methodological approaches were used.

1. Semi-directive interviews with employees themselves;[8] three series of interviews (in 1992, 1993, and 1994) with representatives of subcontractor company management, permanent EDF nuclear power plant employees, contractor (i.e., EDF) doctors and trade-unionists, members of plant CHS-CTs, and doctors responsible for following subcontractor companies' employees;
2. An analysis of legislation and practices relating to radiation protection of workers (i.e., within the plant). This made it possible to identify rules and norms and compare them to actual practices and changes in practices. Between the law and the way the law is implemented or practiced there is always room for interpretation, and this is what makes sense of existent rules in relation to motives and aims that may well be different from those in the interests of which those rules were initially constructed;
3. A study of documents on the maintenance policies of the contractor EDF and the monitoring and oversight policies of workers safety authorities, particularly how those policies were implemented during "incidents" that occurred in French nuclear plants between 1995 and 1999. This study is supplemented by interviews with persons who were either directly or indirectly involved in the incidents.

[8]All first names used here were made up.

The first part of this study presents how maintenance work in the nuclear industry is organized, on the basis of what "outside" workers say of it. The second presents the threefold process—quality, general nuclear safety (against accidents and dissemination of radiation among the civilian population), and protection against radiation for workers—used to legitimate subcontracting of maintenance operations. The third part works to identify the repercussions of this type of social organization of labor on public health; i.e., 1) the health of workers charged with performing maintenance tasks and 2) protection from radiation dissemination for populations living near nuclear power plants.

THE STUDY

Between 1988 and 1994 I met repeatedly with 35 DATR employees, 15 of whom worked on tubes, valves and taps, 6 in non-destructive inspection, 6 in "nuclear servitude" (term explained further on), 4 in insulation/lagging, 2 in electricity, 1 in scaffolding, 1 in painting. Most of them had been temporary workers (either with a limited-time contract or working for a temporary agency) before being hired on a permanent contract. Some remained temporary workers three years into the study or only had contracts on a job-by-job basis that might or might not be renewed. Several interviews took place in respondents' homes, which enabled me to collect accounts from their wives as well. I also met with the widow of a DATR employee who died in 1992. Interviews were supplemented with a follow-up survey questionnaire distributed to DATR employees who had participated in the 1988 study (Thébaud-Mony et al., 1992) and to company doctors. I was also able to attend employee training courses on risk prevention (PR) and the quality-safety training program for "service company" employees (Qualité-Sûreté Prestataires or QSP), compulsory training for anyone working in a nuclear plant, as well as a conference in March 1995 organized by the Conseil Supérieur Consultatif des Comités Mixtes à la Production EDF[9] on "living and working conditions in nuclear maintenance." In recent years, interviews have been conducted with "outside" workers affected by radiation and contamination incidents.

[9] This is a joint "labor council" run by both trade union and management representatives.

ANALYSIS OF LEGISLATION AND IN-HOUSE RADIATION PROTECTION

The information analyzed comes from a variety of sources:

- participation in two public hearings and a taskforce (1998-1999) of the Office Parlementaire d'Évaluation des Choix Scientifiques et Technologiques [Parliamentary office for the assessment of scientific & technological choices] on the issues of radiation protection within nuclear plants and health;
- diverse documentation on regulations, legislation, organization of radiation protection and nuclear workers safety;
- interviews with engineers from the elite training institution of the Ecole des Mines and work inspectors, members of the Office de Protection contre les Rayonnements Ionisants (OPRI) and the Institut de Protection et de Sûreté Nucléaire (IPSN) [Public institutions of control & research on radiation protection], as well as union representatives and occupational doctors;
- consultation of files of the Commission Nationale de l'Informatique et des Libertés (National Commission for Electronic Information and Personal Freedoms) on opening computerized files on ionizing radiation doses received by outside workers.

PART I

Working Under the Threat of Exposure to Ionizing Radiation

More than 80% of nuclear plant maintenance operations are done by regularly employed workers of outside companies that are either direct nuclear industry service providers (including "constructors") or companies that have been subcontracted by such providers. The task of EDF operators is to prepare and monitor maintenance operations; the outside workers are the ones who actually perform those operations on fundamental nuclear plant components. As these outside workers tell it, their work is socially organized around two main dimensions:

- the risk of radiation exposure, which determines both division of labor among the employees of the various companies (owner-operator, constructor, other service providers), and the individual and collective latitude that workers have for managing that risk while performing the operations they are responsible for;
- seasonal plant unit shutdowns that altogether take up between 6 to 7 months a year. This goes together with the "flexibilization" of work, work space, and working hours, thereby subjecting the entire personal, family, and social life of workers to the demands and constraints of nuclear industry maintenance.

DIRECTLY ASSIGNED TO WORK THAT EXPOSES THE WORKER TO RADIATION

Ionizing radiation is among the risks for which the French *Code du Travail* (labor laws and regulations) establishes strict regulation limits, and conditions for ensuring compliance with those limits. The decree relative to protecting workers against ionizing radiation (Decret 86-1103 of October 2, 1986) defines workers so exposed, or DATRs, as "persons whose ordinary work conditions may expose them to more than three-tenths of maximum annual exposure" (Article 3). Exposure of such persons is to be "monitored individually" (Article 34) and they are to undergo "a medical examination" at least every 6 months (Article 36).

In the ordinary language of the nuclear industry social world, the acronym DATR designates "outside workers" who have received the training in radiation protection that is compulsory for all employees of service provider companies working for EDF in nuclear plant "controlled access zones" (i.e., anywhere that radioactive sources may expose them to ionizing radiation). DATRs say that "being a DATR" is a qualification in itself. Workers become certified DATRs by passing tests on material learned in the training programs; they must also pass the biannual medical exam with the company doctor and be certified "fit"—or rather "not unfit for work that exposes to radiation." Given the selective effect of this examination for employees, being certified "not unfit" is, in their understanding, the equivalent of a qualification.

WHO ARE "OUTSIDE" DATR WORKERS?

Today's DATR workers live in the many regions of France where a nuclear power plant represents a major part of economic activity. There are 58 reactors on 20 different sites. DATR workers may also come from other regions in France. They are employees of companies specialized in mechanics, construction, electricity, electronics, lagging, insulation, cleaning. They may also be temporary agency workers who are sent in to meet employer demand, in some cases for no more than a week at a time. These workers show no cultural or occupational specificity. They may be said to illustrate the rise in number of "all-around" mobile French workers, adapting themselves to the company and working conditions of whatever job they have managed to obtain, for a few days, weeks, months, years, between two periods of more or less prolonged unemployment. This is the "new workforce" (or proletariat?) that has come out of 20 years of "flexibility" in the production sector and the general process of socio-economic vulnerabilization affecting individuals (Appay & Thébaud-Mony, 1997; Gorz, 1997).

The workers surveyed for this study were young: their average age at the time the study was conducted was 31.[1] They were living in more than 28 different *départements* across France, some in rural zones around the nuclear plants.

[1] In the STED [Sous-traitance, EDF, DATR] survey done five years after this one, the average age was almost the same: 33.9 (Doniol-Shaw et al., 1995). Five years later, the second stage of the STED survey, during which "outside" nuclear industry workers were again requested to answer a questionnaire, provided supplementary information on how the age structure for this population is evolving (Doniol-Shaw et al., 2001). The average age is the same: 33.8. This means there has been constant turnover of outside workers, the oldest workers having been either fired or dismissed, or excluded from employment in nuclear power plants.

FLEXIBILITY

The work of "outside" DATR workers is structured by one strong constraint. Because maintenance is absolutely necessary to guarantee the operation of a nuclear plant, maintenance work is an integral part of nuclear production. However, maintenance work can only be performed when the nuclear reactor is shut down. And it is imperative that shutdown time be as short as possible. This time constraint has given rise to flexibility—of hours, work, and hiring. This constraint intensified after 1992. The labor of maintenance workers is structured by the reduction of shutdown time and the fact that shutdowns are now concentrated within an increasingly short period of the year: from April to September. This subjects workers to a three-fold process of "flexibilization": hiring flexibility, flexibility with regard to work hours and free time, flexibility as it affects family life.

THE SOCIAL DIVISION OF LABOR AND RADIATION DOSE

Moreover, for DATR employees of "outside" companies, the experience of working in the "controlled zone" when performing nuclear maintenance operations during unit shutdowns crystallizes around the issue of radiation risk. I hypothesized that for work activity, the sense of this risk tends to become the structuring principle of the experience of persons performing these tasks. And indeed, it is within this sense of the work, in intaglio as it were, that the content of the work, the tasks to be performed, come into view; particularly those tasks that subject the worker to high exposure, as well as the rules that frame the tasks and delimit workers' latitude for coping with the threat of radiation exposure and the consequences of actually being exposed. Workers mention the training they receive, the rules and regulations, and what is involved in radiation protection. It is also on the basis of the radiation exposure constraint that they explain how they construct their know-how and their understanding of the other types of risks associated with their activity. Lastly, workers bring to light the social relations that are constructed on nuclear sites, social relations run through with contradictory dynamics, sometimes of cooperation, but often of oversight, rivalry, and competition—and this holds for relations between EDF "agents" (i.e., permanent employees of the owner company and principle) and "outside" workers, but also for relations among outside workers themselves. The way exposure to ionizing radiation is managed is central to construction of these social relations.

SELDOM ASKED TO SPEAK

The experts who analyze industrial accidents retrospectively are highly unlikely to ask operators in charge of daily activities in high-risk systems to recount their experience. The work of these operators and what they might have to say about it remains outside the field of such analysis (Dwyer, 1991; LLory, 1996; Perrow, 1984).

Here, before giving the floor to "outside" DATR workers, it should be underscored that they are *not* usually asked to speak. Comments by French-speaking workers are available, however, in a full-length film by Catherine Pozzo di Borgo (1996), a televised documentary,[2] and a few brief news-report interviews. Workers were only turned to as a group for the first time after a few recent conflicts, most of them quite brief and with very specific focuses. The fact is that both in terms of health and workers' relation to the work experience, what these workers have to say is crucial to developing sociological knowledge of health in the workplace for the activity of nuclear plant maintenance.

After remarking the perpetual denial in Taylorist industrial work organization of what workers might have to say and indeed the very idea that they might have anything to say at all, Philippe Zarifian has shown that their speech is a "discourse about events." What workers reproduce in their accounts is a daily history of experiences of facts and events, practices, chance occurrences, incidents and accidents, a history that is specific to each individual: "The work is made up of a thousand variations, of constant variability (chance occurrences, breakdowns, quality problems, missing material, an absent person, etc.) of the sort that work prescribers and prescribed work take no note of. As norm and as the law of authority, standard language stands opposed to the emergence of such a discourse on events. It ensures that this discourse is perpetually repressed" (Zarifian, 1996, p. 37).

In the world of the nuclear industry, workers' words, their accounts, their discourse are not recognized, nor do their words have any legitimate existence. And yet it is what they have to say that enables us to know about how maintenance work organization is lived in reality, and it is this in turn that enables us to see the forms that social relations of domination take in this type of social organization. The workers' word is both submissive and subversive, because it exposes this kind of work organization itself as a power system, a system of exploitation in which life, death, health, and dignity are at stake.

There are three chapters to this first part of the book. The first presents three of the jobs for which there is proportionally the greatest number of openings in nuclear plant maintenance. They are also among the jobs where exposure to ionizing radiation is highest. The second chapter purports to show how the social

[2]"La Marche du Siècle," June 18, 1997, program headed by Jean-Marie Cavada, France 3 Television, Paris.

division of labor on the one hand, exposure to ionizing radiation on the other, fit together in maintenance activity—I have called this "the social division of the dose." It also looks into what latitude or maneuvering room "outside" DATR workers have for managing the exposure threat by place occupied in the division of labor system and by the constraints of the type of work organization that their activity is caught up in. The third chapter presents the three-fold "flexibility" that characterizes the lives of these worker.

CHAPTER 1

Three Nuclear Industry Maintenance Jobs

> What's a nuclear power plant? It's thousands of kilometers of pipes/piping/tubes/tubing and electric cables and more than ten thousand valves per unit.

This is how a valve maintenance specialist summed up a nuclear energy production center in terms of his own work. Nuclear facility maintenance of this sort consists largely in checking the condition of the pipes, valves, taps and stopcocks that regulate the circuits, repairing anything that is worn out, cracked, defective. The older the nuclear plant the more intense the radioactivity of the system components to be inspected or repaired. The secondary circuit is supposed to be perfectly impermeable to the primary circuit,[1] so work on the secondary one does not officially expose workers to ionizing radiation. But according to testimony from workers I spoke with, in the secondary circuits of 900-megawatt nuclear plant units there is some radioactive contamination. These complex systems of conduits and pipes are held in place by supporting frames, which must also be maintained.

Nuclear maintenance work is broken down into several types of specialized, hierarchically ordered jobs in connection with the part of the system to be operated on. To simplify, maintenance operations fall into one of two categories:

1. The first covers expert assessment and repair of components that are "important for the general safety"—officially listed as IPS[2] procedures. These activities primarily involve tube, tap and valve mechanics and lathing, checking related components through a procedure called "non

[1]The main components of a nuclear reactor are the "core," which contains the fuel; two cooling circuits, one called the primary circuit, which cools the core, the other called the secondary circuit, which cools the primary circuit; and a steam generator. There is also a "water tank" for stocking irradiated fuel during unit shutdowns; the water works as a screen against the radiation emanating from the fuel.

[2]"Important pour la sûreté"

destructive inspection," and electricity. I met with tube and valve specialists and "non-destructive" inspectors; it is their experience that will be presented here.

2. The second category covers the following tasks: decontamination, cleaning, guarding, scaffold-mounting, airlock-mounting, radiation protection, installing and removing lagging materials—everything designated "services." These activities will be illustrated in the section on what is called "nuclear servitude."

VALVE-AND-PIPE MECHANICS

Operations calling for "valve and pipe" mechanics are very important, first of all numerically: it is estimated that nearly 800 outside companies and more than 11,000 outside company employees are involved.[3] These operations are also important for nuclear installation safety, since the quality of maintenance operations insures plant reliability. Lastly, it is important for maintenance worker safety because it presupposes operating on components of the nuclear production system, some of which are highly radioactive, such as the reactor tank, steam generators, and primary circuit pipes, which circulate radioactive water.

Work Content

Opening steam generators, valves and stopcocks; checking their condition and that of seals and soldering/welding; changing certain systems when defects are detected; replacing worn parts, machining damaged or scratched ones—these are the productive tasks of valve-and-pipe mechanics. The examples below will give a clearer idea of them. Some are highly technical and involve moderate exposure to ionizing radiation. For others, the opposite holds.

Highly Technical Tasks Involving Moderate Level of Exposure to Ionizing Radiation

The operation presented by this mechanic concerns a component of the nuclear plant that is particularly "important for the general safety": the primary circuit, in direct contact with the reactor core:

[3]These overall figures are inexact. Given market contract arrangements among the various companies, the sub-subcontracting that occurs, and operation modes, we have no realistic count of outside DATR workers per activity at any given moment nor a clear idea of how the numbers have evolved over time. The only information source is the system for recording radiation data. According to the OPRI [Office pour la Protection contre les Rayonnements Ionisants] annual report for 1997, the number of "outside" workers operating on French nuclear industry sites (EDF, CEA, and COGEMA) that year was 27,061.

> I can cite an example of a "Case 1" modification,[4] an intervention for A. [a sub-subcontracted job performed for a constructor] that consists in dismantling the "ace square" valve in the 1300 MW, and the RRA valve.[5] You dismantle the whole valve. You've got four days. You take out all the screws to remove the valve cover, you remove all the washers—you've got to get rid of the washers. Then you have to bore a hole in the arcade and the cap. They realized that two screw heads had come loose due to the vibrations, which means if you leave it like that there'd be a risk of leakage. The constructor decided to get rid of all the washers and replace them with tightening torques. These are really important valves. You can't run the risk of them breaking down at a crucial moment. They're the valves for the reactor building, at the base of the steam generators. (Guy, permanent-contract mechanic with an outside company, age 35)

Didier, age 26, another permanent-contract mechanic with an outside company, cites other examples of valve work, here a Case 2 that involved part machining:

> EDF decided to change the RRA valve seals but the new ones aren't the same thickness. So they asked us to re-machine the thread and to give it a certain dimension. So EDF gave us the procedure specifications. We get there with EDF equipment. Everything's EDF. If we have a problem, we don't do the operation until they've made the part themselves so we can do the work.
>
> If you've got scratched seal threading or if there's corrosion or whatever, we've got to bring it back to the previous dimension. We do that with a portable mini-lathe. It's EDF material. Everything's set up on a dollycart. We bring it in, set the machine, then we machine it. These operations don't take a long time.

Didier underscores the precision required for these tasks: every scratch, every "jamming" problem has to be dealt with:

> When there's a scratch on the lower portion of the valve, we plane away the metal so it will be smooth and without a scratch, so there's no leaking. Also, for example, they noticed that the valve lids, when we put them back on to valve, there wasn't enough play and it jammed when they were dismantled. So we ran them through the lathe to get it right. We work on all the circuits, even on the flanges at the bottom of the tank, on the SGs (steam generators), RRAs (shutdown cooling system) RCPs (primary cooling system), RISs (safety injection system)—everything in the controlled zone.

It is not just pipes, valves, and stopcocks that must be inspected; the framework holding them in place must also be checked, with the understanding that

[4]"Case 1" refers to operations handled exclusively by the service provider company; i.e., service provider is in charge of both acts performed and equipment. Case 2 and Case 3 operations are performed with EDF equipment and in accordance with procedure specifications established by EDF.

[5]"Ace square" refers to the square formed by the four primary circuit valves—highly radioactive components. The RRA valve is the valve on a temporarily shut-down cooling circuit.

defects may have developed during use; i.e., flaws in the material or work done on it, or unanticipated (unpredictable?) reactions of the material that appear or occur while the plant is in operation.

> Here we're repairing the defective support frame that was made during plant construction. . . . We're redoing the construction work. It was badly done, it has to be redone, but it's not clear how. Because when construction did it, there was no danger. Now we have to spend hours on it. The plant has been working for ten years with badly made circuits. So there are places where it really spits,[6] and we get dosed. (Alain, mechanic with an outside company, 26)

Tasks Requiring Little Technical Work but Involving Exposure to High Levels of Radiation

> Opening the tanks—that's manual work. We remove everything that's above the reactor. We work on the concrete slab that's 20 meters up; the floor is 20 meters below. It takes four days. Afterward they give us side jobs in the shop. There are 58 studs to clean. The studs—that's hardware mechanics; they're two meters high. Instead of forcing the bolts we force the studs. After that we close it up again. In Dampierre, where I just did a tank opening, I got 250 millirems in 4 days [1/8 of the maximum annual dose[7]]. We were supposed to start closing up the day before yesterday, but there are problems, cracked cladding.[8] It's due to oxygenation of the pipes. At least that's what they tell us—they don't tell us everything. There are a lot of radioactive particles. They called G. company to clean the tank walls and when it's gone down we'll be able to finish the work. But there'll be several of us, so we can take over from each other. In any case, we refuse to do the work when the contamination is too high.[9] It's not physically hard work. On sites like Dampierre we work in "Muru" suits.[10] Otherwise we just wear the supplementary outfit. There are four companies in France

[6]"Spitting" refers to a strong, sudden radiation dose on the immediate worksite.

[7]I.e., maximum dose according to the most recent European Union basic norm directive. France has repeatedly had to adapt its regulations to these directives.

[8]This refers to cracks in fuel rod cladding, which means that radioactive particles and gas get into primary circuit water, which then deposits them on all primary circuit components. When the circuit dries, the particles become radioactive dust.

[9]External contamination is eliminated by showers and brushing (done in the medical facility before the worker leaves the plant). Internal contamination requires medical surveillance. Any degree of contamination, even low, causes high anxiety, not only for the "contaminated" person but also their family and friends. EDF has tried to modify the psychological impact of contamination by using another word for it. (On contamination risk regulations, see chapter 5).

[10]Reference is to Mururoa [atoll in French Polynesia, well-known as a nuclear testing site]. What the workers call a "Muru" suit ("bubble suit") is completely airtight, requiring an oxygenation system that allows the operator to breath air coming in from an uncontaminated area. Operations performed in this protective suit are experienced as difficult and "stressful," since it is understood that while the suit protects against radioactive "contamination" (contact with radioactive dust), it does not protect against irradiation.

> that do tank openings. (Eric, permanent-contract mechanic with an outside company, 22)

When steam generators are opened to insert "SG plugs,"[11] exposure to ionizing radiation is intense. Task time is broken up among several operators, each of whom has a precise, short task to perform. Operators are trained to execute the task fast, and fear of radiation is not felt during simulation exercises, which the whole action can seem almost like an athletic performance. Fear comes when the danger of exposure is real, a fear that some workers are unable to overcome:

> Valve specialists open and close the SGs. The opening . . . there's a big buffer, four or five centimeters thick. It's got to be removed with a special machine. What's left is a small plug—that's taken out by the guys who do the SG plugs. It's a small plate, 1 centimeter, with three screws that you can remove relatively quickly. For the opening, we don't need the Muru outfit. It's inside that you get contaminated. Jumpers insert and take out SG plugs. They've got fifteen days of mock-up. They train for fifteen days because you've got to do it fast. Once they're inside, the operation lasts three or four minutes. They're four of them, each with his little thing to do. In three, four minutes, they get 200, 300 millirems [between 1/4 and 1/5 of the maximum annual dose]. But we don't do that. Personally, I did it twice, in '85 and '86, to see how it worked, but I haven't done it since. . . . There are some guys who do the mock-up but then when it comes time to do it for real they don't want to go in. That happens. In that case the guy doesn't do it. They don't force you. (Karim, permanent-contract mechanic, 32)

When the radiation dose is sure to be strong, the team itself divides up the tasks and exposure time among its members:

> An SG manhole—those are operations with a strong dose discharge. It's done in two hours. We manage our thing pretty well. Generally there are three of us. Three of us for setting up the machine; afterwards it's just one. The two others wait outside and after a certain amount of time has gone by we switch: one person sets the machine, the other machines. Then the three of us go in just to take out the machine. Afterwards the third person goes in to take out all the equipment. (Didier, permanent-contract mechanic, 26)

Many workers dread certain operations because they are extremely dangerous:

> I don't like it, it's too dangerous! These are plugs that come in three parts that you put into the primary circuit pipes that come directly out of the core and go through the steam generators, so it's directly linked to the primary circuit. These plugs we put in inside the SGs, we do it for later, when

[11] Plugs inserted into certain **tubes** to stop water circulation during maintenance work. The plugs are inserted before SG tubes can be inspected. Meanwhile, spent fuel is unloaded into the tank—this reduces shutdown time.

> they fill the tank to unload [the spent nuclear fuel]. They're to prevent the water from getting through. While they're unloading, the tank is full, and there are companies that come in to check the tubes inside the SGs. That's why they put in a plug—to hold the water back.
>
> And why is the dose danger particularly high?
>
> Because it's the primary circuit. Where the radioactive water goes through. It deposits radioactive particles on the metal. You've got to work really fast. Usually there are three of us who share the dose between us—and we have to work really fast. You can't panic, because if you panic you just make a total mess of things for everyone. The first time I did it, there must've been three of us and then all of sudden we were just two. Not the same. Instead of going in once, you go in twice! So if everyone chickens out at the last minute, there's no one to do the job. So I couldn't chicken out. I was working directly for that company. And really, it's not the sort of thing you can do. . . . We're aware of the problem. We practice it on a model for a week, we train, we're timed, and then we're just sent in to do it. So we're aware of the danger. But it's a thing you don't want to do more than once in a lifetime! Yet there are guys who do it all the time. (Philippe, permanent-contract mechanic, 22)

A "jumper"—a worker who trains on the model for fifteen days as for a commando operation—found himself up against a difficulty that would have been ordinary for this type of maintenance activity if there were no risk of radiation: the screws got stuck, making it impossible to insert the plug. He had to handle the difficulty simultaneously with the fear of taking a higher dose. Meanwhile the project supervisor outside was pressuring him to get the job done:

> It was worse than usual. It was going badly. I go into the water tank, he hands me the plug, I screw it in on the left side, I screw in all the screws on the other side but one doesn't grip. You can imagine my discomfort. I forced it, then took the whole thing all apart, then put it back in place. And all that time the meter's turning. When I came out I had 197 millirems. "I've had it!" I say, "The thing won't grip." He [the supervisor] says, "What do you mean, it won't grip?" "You can go see for yourself!" I say. "Calm down," he says, "if it won't grip, I believe you." "Yeah," I say, "it won't grip." "Wait a bit, then you'll go back in," he says. "Yeah and what about my dose? You're not the one getting it," I say. Then we talk, all that. I say, "I'm going back, I'm going to get it into place." I go back in, I unscrew the plug, I throw it outside and come out myself. I'm at 200 something. I look at the screw: perfect! Then we take a nut, I put the screw in-perfect. "What the hell!?!" I say. I don't get it. I go back in, he hands me the plug, I put it in. They all take—except two! "I don't believe this!" I say. I see my thing, it's up to 350 millirems! [1/6 of the maximum annual dose]. "I've got to get out of here," I say, "enough is enough." I come out. "There are just two more," I say. "That's better," he says. "I'm going to send in another one of you guys." And the other guy gets the two bolts in! Just like that! He's saying,

> "It won't take . . ." then, "Ah, there it goes!" The second went in perfectly, but we got too high a dose. There were five of us and we got 1.5 rems. We got too much. Compared to the other teams, we were at least 200 millirems above them. [. . .] I took 355 millirems, I think. Once again, too high a dose. Since I already had 1.7 rems, you see, that meant I had 2 rems and something. If the law had passed,[12] I would have been out to pasture. It's too much." (Pascal, temporary contract mechanic, 22)[13]

To be "out to pasture" means prohibited from entering the controlled zone of the plant because of one has exceeded the regulation cumulative dose, i.e., exhausted one's "radiation allowance." This leads either to total unemployment or, for permanent contract workers, partial unemployment. Companies tend not to shift "specialized DATR" workers back and forth between control zone and non-control zone activities. DATR workers, including temporary ones, are thus "kept in reserve" until they can once again perform radiation-exposing work. Since the cost of the compulsory training that EDF requires for these employees is paid by their companies, those companies, including temporary agencies, do not assign these workers to other types of project sites until the diversity of their company activities permits them to. In tube and valve work, this seems seldom to be the case.

Tube and valve workers worry about operating in 900 MW reactor units not only because the danger of radiation is high, but also because of the technical problems of doing the work, which they know is essential to performing other maintenance operations further down the line. In fact, they are assigned to high-dose operations that enable others to carry out different types of operations:

> Usually when we do the SGs, and especially in Bugey and Fessenheim, the oldest plants, those SGs put out a lot of radiation. You've got to expect an average of 1.5 rems for the operation [3/4 of the maximum annual dose]; that is, over a period of fifteen days to three weeks. But when you go into the SG itself, you expect to come out again in the space of three minutes. And you take that much in three minutes. There are hot spots in the SGs that measure 15 or 20 rems.[14]
>
> And what do you go in there to do?
>
> To put in "plugs" that will block the pipes, which is what allows them to work afterwards, move the water in the tank, load and unload [the nuclear

[12] Reference is to the attempt to adapt French regulations to the new European Union directive norms, which lowered the annual limit to 2 rems (or 20mSv).

[13] Excerpt from an interview in the film *Arrêt de tranche: les trimardeurs du nucléaire* [Unit shutdown: the Hobos of the nuclear industry] (C. Pozzo di Borgo, 1996).

[14] Reference is to radiation doses emitted by radioactive sources on the operation site, not doses received by workers.

> fuel] without the pipes behind getting filled with water. Which means the people waiting behind us can work.
>
> There's no machine for opening the SGs and inserting the plugs?
>
> They tried to develop robots but it isn't easy. But it gives us problems really often, especially in the Bugey and Fessenheim plants. They've got lots of things plugged up. They're what you call "leaking" SGs. In the places we insert the plugs, the threading is more or less worn out. Really, they should change them. In Dampierre, they changed the SGs completely. But here that's a problem for them. (Jean-Louis, permanent-contract mechanic with an outside company, 37)

This can lead all the way to an "incident":

> There's one guy who was in the hole for three hours. He got irradiated. . . . He was having a heart problem and it was impossible to get him out. They broke his arm getting him out. (Patrice, permanent-contract mechanic with an outside company, 27)

It was not possible to obtain more information on this work accident, which concerned an employee of another outside company. Patrice witnessed what happened but could not learn anything more, since his company was not directly concerned and left the plant that very day.

Other operations where radiation risk is high have more to do with cleaning and do not prepare the way for crucial operations such as loading and unloading nuclear fuel. In 900 MW plants, considered by DATR workers the most seriously contaminated, radiation and contamination risk are very high and operation conditions extremely difficult.

> I restart the SGs. That means removing the deposits ["mud"] from the secondary side of the steam generator and doing the rinsings. There's not much contamination because the surfaces are cleaned beforehand, but there's strong radiation. We work on gratings above highly irradiated mud. That's a particularity of 900 MW plants. The circuits are narrow, there's very little space for passing tools and equipment and it's hard to move. All 900 MWs are "rotten" [i.e., highly contaminated with radioactive particles]. (Gaëtan, permanent-contract milling-fitter "leased" back and forth between two outside companies, 26)

Valve and tube workers' activity in nuclear power plant controlled zones is constantly dominated by two types of concerns, one involving the characteristics and constraints of the task itself, whether technical, organizational, or concerning hours worked; the other tied to the need to limit as much as possible their individual exposure to radiation and contamination.

Constructing Qualification: What Training, What Work Recognition, What Recognition for Know-How and Certified/Skills?

Compulsory Training Programs

Before being allowed to enter a nuclear plant, all outside company employees must pass a compulsory EDF two-part training program: "risk prevention" (known as PR) and "service provider quality and safety" (QSP).[15] After the two brief training periods (usually each a week long), trainees must take a test to show satisfactory acquisition of program information and knowledge.

> Pretty soon you're going to have to have graduated from a "grande école" [France's elite educational and professional training institutions] to set foot in a nuclear plant—given all they ask from us. I mean, taken all altogether. The QSP is everything about how the plant functions, all the material used in IPS, worker safety, all that—that's the whole QSP. They talk a little about everything. Then there's all the papers and jive you've got to fill out, all the papers. They keep requiring more and more. There's more than the QSP. You have to take the HNs, the RP2s,[16] get the certificates. For all the monitoring devices. You've got to know how to note doses, how to note everything that's radioactivity. But you've got to manage to learn all that. (Alain, permanent contract mechanic, 26)

The tests given at the end of the training programs cause some anxiety for DATR workers:

> I get stressed for the exam, I mean when we've got a test to take. I've got the commute, the exam, I've got to find a spot for my trailer . . . you don't know what they're going to ask you—it makes you really anxious. (Jean-Louis, mechanic, outside company, 37)

They perceive the training as a selection instrument:

> I pity foreigners who don't know how to talk French or who half understand, and they've got to learn it all and they've got to take the exams. I don't know how they do it. Because they can't have an interpreter, it's the guy himself who's got to answer the questions. It's not easy. And it seems to me they're asking for more and more. Maybe it's so they can choose which ones to take, to eliminate more and more of them, eliminate them from working—maybe. I don't know. Maybe EDF wants to select an elite to work in their nuclear power plants. (Alain, permanent contract mechanic, 26)

[15] See chapter 4 for a detailed presentation of training programs and organization.

[16] HN: "*habilitation nucléaire,*" certifying fitness to work in a nuclear power plant; RP2: advanced risk prevention.

> The QSP? Didn't teach me a thing. It just allowed them to eliminate plenty of people—guys with five, six years on the job. (Patrice, permanent-contract mechanic with an outside company, 27)

Skills and Certification that Permanent-Contract Employees Can Never Definitively Acquire

The technical skills involved in tube and valve work are mechanics skills. Permanent-contract employees working on IPS material already necessarily have a degree in mechanics, in some cases also in maintenance and automation (French vocational school degrees in mechanics; a vocational *baccalauréat* in industrial maintenance with a specialization in nuclear maintenance). There is no specific initial training for pipe and valve work. Some have done a two-year technical program after the baccalauréat (bac + 2), obtaining a degree in a physics-related field from an Institut Universitaire de Technologie; others are at the bac + 5 level (one young man had studied biology and obtained a DEA [pre-doctoral] degree). They were unable to find jobs at their qualification level.

Respondents say that after the training they learn from experience:

> I learned with the veterans. Pipes and valves—that doesn't exist. I mean there are no programs for training pipe and valve specialists. We're trained by the company, and then the company sends us to train at the plant for three, four, six months, with revision courses often because it's very, very important. It's a pretty odd profession, actually. It's mechanics if you like. It's huge valves that we put on and take off. You can't make any mistakes. A nuclear power plant is not a factory. That's why we've got a specialty—valve mechanic—and that's all we do. I've been project leader for around two years. We've got responsibilities that are actually pretty heavy. When we come on the job, we work with what's called a "régime de consignation"[17] and those papers are in your name. If a mistake is made in what you did, they pull out your name and the name of the company. (Karim, permanent-contract mechanic with an outside company 32)

> I was a car mechanic. I applied for a job here because valves, that's mechanics. I learned some things on the job. I've been with machiners in the nuclear industry for three years. I was a subordinate for two years. I learned well. By watching at first, by learning, by asking questions and later by taking the machine, always backed up by someone else, and now I do everything myself. You can't learn all those things at school. There's the operation itself. You're not going to be practicing on a life-size SG. It's a whole field, the SAS SG. The mock-up SG is not really the same. (Didier, permanent-contract mechanic with an outside company, 26)

[17] A specific prescribed procedure for each individual maintenance operation.

In addition to the initial training and experience, there are "nuclear accreditation" programs (HNs; see n. 15) that have to be periodically retaken as they are compulsory for continued employment in nuclear power plants:

> The accreditations are to know your level, to know what task you can do. As a subordinate (HN1), there are papers I can't do. It's really complicated. I should've brought you the book they give you because I did a program this week. It's got all the accreditations. The HNs are all the different levels. We do a training program, an HN program, in the company. That explains the safety you've got to do, what you're allowed to do and what you're not. It's the employer who delivers the HN, he's the one who determines the level. Obviously, to get the accreditation, you've already got to be a DATR. There are three levels: HN1, HN2, and HN3.
>
> And an HN3 has greater responsibilities on the project?
>
> It all depends. Sometimes you can be HN3 and a subordinate. It's a question of price. If they want three HN3s to do a valve, they'll pay them more than three HN1s. But usually the HN3s are project bosses. (Eric, permanent contract mechanic with an outside company, 22)

Limited Training for Temporary Workers

For workers who arrive at the job through a temporary agency and who have to perform operations that are "dose-costly" but do not require particular technical skills, training is minimal. A request for complementary training by a DATR temporary worker (with a vocational degree in metal fitting mechanics) who had already done the basic EDF compulsory program (DATR and QSP) was turned down by the temp agency:

> I felt limited when it came to valve operations, and I wanted to take one of the training programs offered in connection with EDF but the employer refused. Once, on the worksite for operations on the primary circuit, there were four of us temp workers, including the project boss, all with no experience in nuclear maintenance. The project boss came from the steel industry. We had to figure out how to manage. (Daniel, sub-subcontracted metal fitterworking for a temp agency, 30)

Despite EDF's training demands, then, the question of qualifications seems elided when temporary workers are called in, since the situation that Daniel describes was possible, a situation in which it seems no one had any experience in "nuclear acts." Daniel expressed the need to acquire training that would enable him to perform more confidently the tasks he is asked to perform, knowing that he could find himself with other temporary workers in the controlled zone with no experience of the work to be done.

Daniel's demand was not taken into consideration by the temporary agency employer. For the agency, complementary training that would increase a temporary worker's qualification level amounted to providing that worker with the means to demand higher pay. Trends in the temporary agency market show that agency profit margins over temp worker wages are falling.[18] Unless a training program is made compulsory by the principal (i.e., EDF), it is not in the temp agency's interest to provide complementary training to its workers. Quality demands are negotiated between EDF and subcontractor companies; it is the latter's responsibility to provide personnel that can satisfy EDF demands. The temp agency as employer does not feel required to provide any more than the basic compulsory training, even though it has committed to meet the accreditation level required by EDF. This explains why the temporary worker's request was turned down. The temporary worker was aware of and anxious about the safety breach resulting from his inexperience and that of his fellows, also temp workers.

Training of "Jumpers"

Workers who perform interventions that expose them to high radiation discharges are called "jumpers." They have to take a supplementary training program that is more like athletic training, aimed at self-control and quick performance. The training is made up of simulation exercises on a model and directly faces the question of the danger the worker is in and the absolute necessity of limiting ionizing radiation exposure time. Jumpers are valued for performing this "athletic champion" type of exercise.

Still, the way the jumpers interviewed perceive their work is very different. They feel they are caught in a system that they could only get out of by means of a permanent contract,[19] as one explains:

> I don't like it—it's too dangerous! I'm going to quit, in any case. If they don't hire me permanently in the company I'm at, I'm leaving. . . .
>
> We'll make less on sites outside the plant, that's for sure, but we won't get the dose. When I hear the guys say, "We can take the dose, it doesn't do nothing," I don't agree. When I've reached my limit, the company's going to say: 'You've gotten enough? Get out!' That's what they're going to do to me, it's no use kidding myself." (Pascal, temporary contract mechanic, 22)[20]

[18] When a temp agency "leases" a temp worker to another company, the cost billed to the latter is calculated by multiplying worker's wage by a coefficient that is more or less standardized at the level of the entire temp work market. The coefficient was above 2 a few years ago; it is now at 1.85 to 1.9—still almost twice as much as the temp worker's wages (Fargeot, 1997).

[19] Temporary agencies promise jumpers that they will be able to perform a high number of operations per year; in fact, they only perform a few, and spend much time at home waiting for work calls.

[20] Excerpt from Pozzo di Borgo's film, *Arrêt de tranche* [Unit shutdown].

Pascal's fear is of course two-fold: it concerns the effects of the radiation on his health—though messages meant to be reassuring are circulated on the reality of this effect—and the worry about reaching "the limit" (i.e., exhausting his radiation "allowance") and consequently losing his job.

Many DATR workers undergo months of unemployment not because they have reached what is considered a maximum dose fixed in order to protect their health from dangerous risk but rather because they have exhausted the "radiation allowance" they are "entitled" to.

Recognition of Know-How, Skills, and a "Return on Experience"?

"Veteran" workers over 30 have ambivalent perceptions of training programs and the tests that certify acquisition of the knowledge acquired in them. On the one hand they find it interesting to acquire knowledge about "the purpose served by the circuits of an operating nuclear plant" through the QSP training programs, since they are always working on these circuits. They are also glad to receive training in worker safety and radiation protection. On the other hand, they worry that the "demonstration of acquisition of theoretical and formal knowledge" requirement that is part of the training is becoming more important for being recognized as qualified to operate in nuclear plants than the practical knowledge they have acquired in the framework of real work as pipe and valve mechanics in those plants:

> You've been doing this a long time. Did the training help you? Do you feel you learned something?
>
> Learn something! You learn something, but there comes a time when you forget it. Because listen, we do a recycling every two years, there are things we never practice. We learn, we learn—you've got to have a head as big as a laundry boiler. But there are things we don't practice enough. We don't get recycled often enough, I don't think. We should get recycled more often than twice a year if we're going to cram all that into our heads. . . . The veterans, they moved up with seniority, experience. In the old days, a guy who started as a manual worker could become site boss or even higher. It's a lot more complicated now, I think, because you've got to redo and redo the training programs constantly to get back up to snuff. They ask for more and more. These days the guy who climbs the highest is the guy who comes out of school with the most baggage. He's maybe even a guy who doesn't know the job as well, but he's got the most baggage. (Alain, permanent-contract mechanic, 26)

What Alain is attesting to here is the process of devaluing the experience and know-how acquired in the activity itself compared to the knowledge taught in the training programs, knowledge transmitted through conventional, normative discourse about the prescribed work.[21] Isn't this also a disguised procedure for evaluating ability to write, given that for EDF this is an indispensable skill because it ensures the traceability of work performed? In fact, the link between maintenance worker subordinates and maintenance operation organizers depends on that traceability.

Whether in training programs or work inspection and evaluation, workers' construction of know—how and concrete experience of the plant and its parts in different modes by type of task performed is not recognized as such, nor is task variability or the specificity of situations encountered taken into account in the way work is organized. Because variability is denied, the know—how that enables workers to handle it is also denied.

"Outside" workers emphasize the increasing dissociation between forms of accreditation (training programs and tests) and the construction of experienced-based know-how. In fact the "return on experience" so often mentioned in training programs or in the discourse of EDF maintenance heads corresponds for the workers to a completely different reality.

In the course of the interviews, the "return on experience" seemed like an obsessive preoccupation for some. It took me some time to understand that the expression did not refer to recognition of maintenance worker skill and know-how. The "return on experience" in EDF's language means not human experience of nuclear plants but 1) optimizing maintenance through optimal use of the information signaled by technical systems on degree of wear or malfunctioning, and 2) keeping written records of work performed, which are required to meet general safety authority traceability demands.

> They didn't use to have the "return on experience." So what they did is they'd open a lot of valves—3000 to 3500 valves during one unit shutdown. Maybe 2000 of them didn't need to be opened. Then over the years they got their return on experience. After [a valve's been] four or five years in operation, they say, "We've opened that valve 5 times, there's never a problem, so next year there's no reason to open it." That limits things. So

[21] While participating in a PR training program, I witnessed a brief exchange between a DATR worker and trainer. The trainer, using an overhead projector and transparencies, had presented the sequence of procedures for a given operation on a specific valve. The DATR worker, who had just performed the operation in a nuclear plant, spoke up: "Last week I had to do that, and in fact it didn't happen like that. . . ." The trainer interrupted him before he could say anything else about his experience: "It's not supposed to happen any other way than the way I told you." This remark demonstrates the denial of variability that is characteristic of the prescriber's way of formalizing the work to be done. Both sociological and ergonomic analysis of work activity show that no operation is standard. Variability is not taken into account either in the procedure specifications for nuclear industry maintenance operations or the length of time scheduled for their performance.

> we've got shorter shutdowns. But there are systematic valves that you're going to have to open all the time. (Guy, permanent-contract mechanic, 35)

> At the beginning they looked at everything. Then if after ten years they realized that this or that valve had never had a problem, they might only open it every two or three years. But for the valves that tend to wear out, they open them every time. (Didier, permanent-contract mechanic, 26)

The technical "return on experience" technique seems dissociated from the human experience of the subcontracted workers and their contribution to the efficiency of more narrowly targeted maintenance. The know-how that allows them to contribute remains invisible. It is as if only the vigilance of "experts," technical processes, and written administrative oversight were responsible for the improved efficiency of maintenance activity. The "return on experience" allows for rationalizing time and content of unit shutdowns, but this is to the detriment of the task performers—i.e., the DATR "outside workers"—because it allows for decreasing the volume of work and "seasonalizing" unit shutdowns. The reduction of unit shutdown time made possible by maintenance rationalization does not lead to recognition of workers' skills and thereby does not lead to promotion and improved wages for outside workers. On the contrary it means intensified work (increased productivity), more systematic use of temporary workers, and falling wages due to longer overall unemployment time.

> When I began, once every ten years the unit was stopped for 95 days. Now we've got ten-year shutdowns of from 27 to 35 days. We just can't do in 35 days what we used to do in 90. (Guy, permanent-contract mechanic with an outside company, 35)

> In those days [the 1980s] we had a lot less work, we had a valve to do per week. Now we really work a lot. But still, they're wouldn't give us 10 valves a day! . . . (Didier, permanent-contract mechanic with an outside company, 35)

> They bring in more people, and there's the return on experience now too. (Guy, permanent-contract mechanic with an outside company, 35)

Sub-Subcontracting as a Means of Inhibiting Recognition of Know-How

In sub-subcontracting, workers for second-level companies explain that they feel less vulnerable with regard to the EDF principle than when they are directly subcontracted by EDF. In exchange, they are aware of losing their occupational identity as employees of their own companies.

> On Case 1 modifications, it's the constructor's equipment. They bring in their own equipment. It's the constructor performing a service for EDF. It's the constructor who contracts with us and they bring their equipment.

> It's the constructor who determines the modification and who imposes it on EDF. The advantage in working with the constructor is that they do "Case 1s." So if there's any problem, they make EDF accept a modification before starting the job. If EDF doesn't want to do it, they won't work for them; they say "We're not going to do the work." If there are no anchor supports for lifing the valve cover, if something's missing, they say, "We're not going to do the work. It's not like we don't have anything else to do." (Guy)

> We only do "Case 2" work. We bring our professional knowledge and they [EDF] give us the specifications. In "Case 1" they [constructor's employees, directly subcontracted by EDF] work with their own documentation. Quality plan, procedure specs, everything. Everything is handled by the constructor. We're not accredited at this time to determine the procedure specs. So we don't change anything. Whether we're working with the constructor or EDF, they're the ones who give the specs. When we work with the constructor, we're with them, we're their guys. We leave the name of our own company at the door and we become the constructor. It's not just a cover. At EDF they know full well that we're from this or that company because the constructor we're working for doesn't have any valve specialists. On site, we're completely the constructor's employees, we sign quality specs for that company. We work on their procedure specs, with their equipment. Vis-à-vis EDF, we're all with the constructor. (Didier)

This apparent assimilating of employees from an outside company to direct employees of the constructor that EDF has hired at the first sub-contracting level means the reality of sub-subcontracting disappears. Because these workers are identified with the constructor, their own professional identity and specific know-how becomes invisible. In the actual activity, they enjoy significant maneuvering room as the equivalent of constructor employees, but the trade-off is that they receive no recognition for their own "expertise."

There is no "return on experience" for the job of nuclear plant valve mechanics and the practical knowledge they have of plant installations. EDF's "paper" monitoring of the content of the operations these workers perform seems to them very remote from the concrete experience that they think of themselves as repositories of. Written information is the main vector linking service providers and EDF work planners, and it must be recalled that this information is noted in accordance with norms that leave no space for an account of operators' experience. The valve maintenance workers I spoke with feel closer to the constructors than to EDF but mention the fact that no one turns to them, even when they have the solution to certain technical problems:

> If you've got a cracked plug on a valve classed "mine" [name of a particular category of plugs] you're not going to take the initiative of deciding how to repair it. No, the valve constructor has got to be informed that there's a crack in his valve. And he's already defending himself. Whereas we could

> immediately give them the solution. We tell them: "You've got to do this." But if they're not the ones holding the papers . . . ! (Guy)

Once one has the papers, several years and numerous negotiation sessions with EDF and the safety authorities have to be gone through before the proposed solution is recognized—perhaps, and no one calls in the person(s) who made the suggestion on the basis of their practical experience or seeks to discuss the question with them. The professionalism and know-how of tube and valve workers are essential to the functioning of EDF's "return on experience" and the quality of nuclear plant maintenance. But the experience and know-how of DATR outside company employees are not given any explicit, formal recognition that would then be reflected in improved status, careers, or pay. With the many levels of sub-contracting, the space of salary negotiations between the principal and work prescriber, EDF, and those who perform the work prescribed disappears.[22] These workers therefore cannot use the value that their professionalism represents for those who benefit directly from it—EDF—to negotiate for recognition of their qualifications and, with that recognition, better working conditions and pay.

On the contrary, EDF appropriates the results of their experience, which allow it to limit maintenance activity and diminish costs to itself, by constraining service providers to comply with the logic of choosing the "least expensive" in the competitive bidding procedures. This mechanism is then reproduced by every service company with regard to the companies it subcontracts. Companies that choose to skill stable employees thereby lose jobs to less expensive companies whose workers are not as well paid or who use temps. And this has happened because the return on human experience has actually meant that there are now procedures for performing certain maintenance operations smoothly and skillfully enough that they can be performed by less experienced personnel. The ability of the employees of each subcontracted company to negotiate hiring, wages, and working conditions is sharply limited by the bidding procedure between principal and first level subcontractor companies. Moreover, it creates competition between employees of all the different subcontracted companies, all governed by different collective labor agreements.[23]

[22] I encountered at least one case of 6-level subcontracting.

[23] In a study of subcontracting in the Chinon nuclear plant, Michel Lallier, secretary of the Confédération Générale du Travail union [CGT, **main French union organization**], counted 46 different collective labor agreements for the 2500 outside workers that had performed maintenance operations in the plant in 1993. Employees at the same qualification level covered by different collective labor agreements may well receive quite different wages. A valve worker covered by a steel industry agreement is paid much better than one covered by the building and public works industry agreement (Lallier, 1995).

NON-DESTRUCTIVE EQUIPMENT INSPECTION SPECIALISTS

These workers call themselves "nuclear plant nurses." Their role is to "auscultate" the different structures composing the plant production system, e.g., steam generator tubes and guide tubes that come off the control panels, to detect cracks and other signs of corrosion or wear that could bring about leakage of radioactive effluents; places where welding joints are giving out; or any other danger of malfunction. Like certain valve-handling tasks, this activity is of major importance for safety and cost: inspectors' "diagnoses" lead to decision-making; namely, whether to leave equipment as is or repair or change it. This in turn means constraints in terms of qualifications and ongoing training for employees and companies.

Though inspection work is not specific to the nuclear industry (inspection is also part of maintenance operations in the petrochemical, aerospace, and building industries as well as in sheet metal work), performing it in nuclear plants presupposes acquisition of specific knowledge through specialized training in addition to the PR and QSP programs.

Until 1994, the relatively high qualification requirement meant there were few employees skilled to do this work. Today there are approximately 1000 inspection technicians and engineers, 200 working directly for EDF while 800 are subcontracting company employees. The latter group used to have some latitude for negotiating with the employer company and they used to be able to move fairly readily from one company to another for this kind of work. Though the contracts were often for fixed duration, there were few inspectors on the job market (once again, because the required qualification level was high), so they were almost automatically rehired by one company or another. This flexibility was threatened in 1994, when international competition for EDF jobs, specifically German and American, threatened sector employers; the sector underwent lay-offs and intense restructuring. Small companies disappeared and employment became unstable; employees observed that the number and frequency of inspections was falling. And like other nuclear plant service provider employees, they were subjected to shorter unit shutdown times.

> The situation in inspection has gotten much worse. There've been a lot of lay-offs in the companies and I don't have any idea what it's all going to come to in the future. It happened when EDF changed policies. They reduced unit shutdown time: shutdowns got shorter and shorter, so operations had to be shorter and shorter, which meant the companies needed less personnel. Inspection companies used to be in a fairly privileged situation, now it's going downhill. There were lay-offs in other companies. There weren't many in ours until now, but it's unstable, insecure. Guys who don't have enough certifications can get pushed out. (Marc, inspector on a fixed-term contract, 33)

Work Content

What does non-destructive inspection consist of? With regard to the activity strictly speaking, different techniques are used by type of inspection to be done: ultrasound, ressuage liquid penetrant inspection, Eddy current, gammagram imaging. Many of the inspectors I interviewed practice several techniques. One explains gammagram imaging inspection:

> Its like a lung x-ray, to see inside. It can detect holes, cracks—the same principle as for photography. The particularity of these EDF inspections is that it's follow-up on the life of the plant. There are regular yearly inspections, and if we find particular flaws, sizeable but not dangerous, they do follow-up on those flaws. All we do is inspect, we don't do the follow-up. We take a picture, at most we interpret it—they handle the rest. . . . The basic instrument is the irridium source. It's in a depleted uranium container. There's a remote control with a ten-meter-long handle. There are other types of instruments that use cobalt and have electric controls. (Marc, inspector, fixed-term contract, 33)

This work involves exposure to both irradiated nuclear plant structures and the radiation source used to do the gammagram imaging. Operating conditions in the nuclear industry are thus perceived as very different from those in other industrial sectors. Dealing with radiation dose is an ineluctable component of these operators' activity.

> I worked in aviation radiography but in a protected area; that is, in a booth where you don't get any radiation except if there's a leak. . . . The working conditions in a nuclear plant are totally different from other sectors because you're working in places where protection can't easily be installed—there's not enough space. No booth. We have fairly long exposure times in the SGs, you can't really protect yourself. We relay each other a bit, we run pretty fast behind a wall. (Jean-Claude, trained as an electrician, fixed-term contract, 25)

Protection measures are uncertain and managed by the inspectors themselves, who seems like the only ones taking any responsibility for their safety.

> Ressuage liquid penetrant inspection is a surface technique: we inspect the seams, the joints on the generators and the pressurizer, some of the secondary circuit valves. This work is not particular to nuclear plants, but in a plant there are additional problems due to where you are. With ressuage, for example, there can be a problem of surface contamination, so we have to use gloves and wear a special outfit. (Marc, inspector, fixed-term contract, 33)

Another characteristic of gammagram imaging inspection is that it must not interfere with activities on other sites, the danger being that EDF workers in the area could be irradiated:

> There are a whole lot of companies working at the same time in the plant, and it's not always clear how to proceed. We don't have that problem in gammagram imaging because we work at night. We can't work at the same time as the others because we work with radioactive sources. The sites are already highly contaminated, and we bring in sources that can get up to 200 curies. So we can't work at the same time as the others. Sometimes we work during the day, but only on weekends. In fact, we always work outside normal production hours. That's the biggest pain, but we don't really have a choice. (Marc, inspector, fixed-term contract, 33)

This remark brings out the strong working-hour constraints inspectors are subjected to. Night inspection results are checked during the day, which in turn allows for organizing work for the coming days and nights. A crucial stage in organizing maintenance work is scheduling time for inspections; the extra time needed for interpretation is not taken into account in real-time scheduling. To reduce reactor shutdown time, the principal has chosen to force workers to accept this shift organization, which implies night work and forces inspection companies to depart from the Sunday day-of-rest rule. This is done by loosely interpreting the notion of "urgent maintenance work," the *Code de Travail* term for situations that allow departures from rules that have not been negotiated as part of union accords.[24]

In response to a question addressed in 1989 by the Direction Régionale de l'Industrie, de la Recherche et de l'Environnement (DRIRE) of the "Centre" region [200km south of Paris] to a bureau of the French Ministry of Labor on the matter of rules for departing from the rules, M. Dutheillet de Lamothe, then director of labor relations, noted as follows: "Application of the special dispensation procedure provided for in Article L 221-10 of the *Code de Travail* is subordinated to intervention from an extensive sector agreement that admits the possibility of organizing continuous work for economic reasons. Only industries in which such an agreement has been passed are permitted to depart from the Sunday day-of-rest rule." This provision does exist in collective accords in the steel industry, but not in most of the many collective accords covering "outside" DATR workers. M. Dutheillet de Lamothe adds: "It seems to me that the relevant companies, which are likely to work during planned and unplanned nuclear plant shutdowns, are covered by the clauses in *Code de Travail* Article L 221-12 that provide for the possibility of suspending weekly rest for the personnel of companies performing urgent operations in another establishment to organize rescue measures or prevent accident damage to equipment, installations, or buildings in that establishment." The notion of safety emergency seems here to have slipped toward a notion of economic emergency. At issue are pre-planned maintenance operations that are not designed to prevent or repair anything in

[24] *Code du Travail,* Article L 221-10.

response to immediate danger but are instead part of permanent, long-term maintenance. The only justification for Sunday work is that it shortens plant shutdown time through a sophisticated optimizing of the work schedule. Moreover, using the "emergency" notion is a way of eliding the requirement that a negotiated accord be reached between employers and union representatives of the employees in question before weekly rest can be suspended.

Constructing Qualification: What Training, What Work Recognition, What Recognition for Know-How and Certified Skills?

Like pipe and valve workers, inspectors have to take PR and QSP training programs. But these are only part of their required training. Specialized inspection training is a major ongoing component of work planning as well as a means of selecting employees and companies.

> Next week I've got training. We've got eight weeks of training to do a year—EDF requires it. It's called adapted training. It's not directly paid for by EDF, but in the billing they compensate with an increase of 30% per employee. It's included in the employee's leasing price. (Antonio, permanent-contract inspector with an outside company, 50)

> For EDF, we have to do three or four weeks of training every year, even me, though I'm a veteran. It used to be paid for by EDF, now it's the company. It's to keep us on top of things every year, because there are lots of procedures that change, work methods too. Inspecting often involves robots, and robot design changes all the time, so to operate them you need training. We have to diversify and keep up on all the procedures.
>
> Does the fact of shorter shutdowns these days affect your work?
>
> Well, instead of having two teams for two months, we've got four teams for one month. And now with the fuel that's only going to get changed every eighteen months, that means a plant shutdown once every eighteen months instead of every twelve, and that's going to give us a lot of problems. We're going to have to spend two years getting up on things. I think they're trying most of all to get the number of contract workers down to a minimum and skill them a hundred percent. Because they used to have a lot of personnel that only had one certificate, in liquid penetrant or magnetic techniques. Whereas now you've got to have at least three "level 2s." And they're also setting up a certification program at the national scale that we're going to have to get to be able to do inspections. So all this is changing. When you start at the plant now, you've got to take exams, because they noticed that there were service companies that didn't know what the work and safety legislation is. So before getting into the plant now, there's half a day of tests, and they're getting harder and harder. Each time you go in there are written multiple choice and essay tests. The guys who fail either take the HN1, that is, Level 1, so that for the time being you can't manage a team, or else they're fired. And that's in addition to the QSP.

> Do all these tests seem useful to you?
>
> In my case, given how long I've been at it, it's starting to get a little annoying. But I put myself in the place of some of these plants. There are service workers sometimes who don't know how to read. You've got to admit, that's kind of dangerous. I think it gives more importance to the exams. We used to know them by heart, it was easy. It ups the value of the entrance examinations a bit more, I think. (Jacky, inspector with an outside company, 50)

Inspectors think it is important for EDF to verify maintenance operators' safety training. Still, for a skilled worker, the repetition these tests involves, especially in a context of unstable employment, and the way the tests are used to select workers, represent a permanent threat of losing their qualifications, while passing the tests and being officially qualified offers little hope of career advancement:

> When its new, the first year, youre interested. Afterwards, you see that you're in a system that doesn't let you get very far ahead and that doesn't necessarily fit you very well. That's the real drag. I'd be happy to work for a company where I could get COFRENDs (certificates that qualify to do inspection). That's the only way to get better pay and more interesting work—though inspecting doesn't interest me at all. I'd rather do something else. Like many people, I didn't choose my work at all, and I find myself stuck on the rails, specially trained . . . like, specialized! (Marc, inspector, fixed-term contract, outside company, 33)

Inspection is nonetheless an essential component of the "return on experience" mentioned above. Every inspection process involves a voluminous file specifying method used, results, and inspection analysts' interpretations of them. And once again, the physical task itself and the know-how it requires remain invisible in subcontract relations; the only trace they leave is the contract signed between the companies, and an account of the work in file form. Only when there is a problem does anyone pay any attention to the signature of the person who coordinated the work and was responsible for the conditions in which it was performed.

As in pipe and valve work, the "return on experience" has been used to reduce inspection activity itself. This approach has made it possible to decrease the number of plant installation inspections and space them more widely. The data collected were not used to reorganize scheduling or modify conditions of exposure so as to reduce radiation doses. Inspectors too are now caught in the vise of unemployment, either partial or total, and their ability to negotiate their spot on the labor market has been weakened. The threat to their jobs from company restructurations has mobilized workers in the biggest companies, where the unions are present. When union representatives were sacked, the employee mobilization was supported by EDF unions and brought to public attention by the media. But the mobilization did not last. Every collective move within this work organization system is ephemeral because of the labor relations specific to subcontracting. In a service-provider-to-client relation (i.e., subcontracted employees and their employers to EDF), there is no space for negotiated regulating of labor and hiring conditions. The transformation of work relations into exclusively market relations, including for matters of health, hygiene, and safety, has thwarted exercise of union rights. In addition to the work scheduling pointed out above, there is no negotiation of the conditions of the operation: protection, equipment, methods, etc. Every attempt to start a labor conflict is disarmed by employers citing the threat of not being rehired by EDF.

This situation has even stronger effects on "nuclear servitude" workers, on the bottom rung of the nuclear industry maintenance activity ladder.

"NUCLEAR SERVITUDE" WORKERS

In the previous presentation of mechanics' and inspectors' jobs, mention was made of the "services" that allow them to perform those jobs. Taken together, these services are called "nuclear servitudes" in France—a highly suggestive term for designating where these workers' activity is positioned in the social division of labor.[25] They are there to make it possible for the other maintenance tasks to be accomplished without the workers who perform them receiving excessive radiation. On-site operations may be divided into phases: the most qualified workers carry out the most technical phase; the others are responsible for preparing the ground for those operations and putting the site back in order afterwards.

Nuclear servitude workers have three functions, as essential as they are complementary: cleaning, removing lagging (that is, exposing pipes and other structures), and radiation protection. In the language of the nuclear industry, cleaning is called "decontaminating"; that is, removing radioactive dust from reactor circuits where it tends to settle. Decontamination is a necessary radiation protection activity, as is construction of lead or vinyl airlocks, suiting and unsuiting surveillance and assistance, and monitoring of workers entering strong radiation emission spaces. If this work were not being done in a nuclear power plant and therefore did not involve exposure to ionizing radiation, it would probably be given to women, as is classically the case in all paid occupational activities that recall domestic work. "We're called spirits of the hearth"[26] said one interviewed worker. However, maximum allowed doses for women are much lower than for men. Existing data, from this study and the STED survey

[25] In the *Dictionnaire Robert* (1982), *servitude* is defined as 1) "slavery," "serfdom," "submission," and "subjection"; 2) that which creates or may create a state of dependency (see also the Glossary).

[26] "Les fées du logis."

(Doniol-Shaw et al., 1995) show that "nuclear servitude" workers are generally young men who agree to work on a fixed-term contract or temp basis.

As in offices and factories, cleaning work in a nuclear power plant is the "reverse" of skilled worker tasks, here maintenance. As Hélène Bretin has shown (1994) for street-cleaning and garbage collecting, this is work that is socially recognized only when it's judged to have been done badly: when dust or, in this case, radioactivity persists. What's at issue in this work is the other workers' safety and protection from the risk of irradiation and contamination. If site dose levels are high due to surface contamination in the form of dust, higher-level maintenance workers criticize cleaning workers. Nuclear servitude workers are on the bottom rung of the labor ladder, the bottom of the pecking order.

Nonetheless, EDF delegates to these workers the task of checking that other DATR "outside" workers follow radiation protection measures in suiting/unsuiting operations and when entering and leaving airlocks. This function positions them paradoxically with respect to their colleagues: they are both socially denigrated workers because of their low place in the division of labor, and in an oversight position which gives them a certain power over the others.

The Activities of "Nuclear Servitude" Workers

A 23-year-old decontaminater who has worked in nuclear plants for four years and had nothing but fixed-term contracts describes his activity thus:

> I do upkeep: cleaning, decontamination, servitude. We get the sites ready for operators, that is we put on lead protections and we try to limit the dose for the operators who come after us, so we're the one's who get it up the nose. We also wait on them: suiting, unsuiting. (Hermano, nuclear servitude worker, fixed-term contract, outside company, 23)

Decontamination

Hermano adds suddenly that servitude work consists primarily in "getting dosed." Decontamination is in fact a particular cleaning task because of the high exposure to irradiation and contamination:

> What do I do? I go where it spits. We're "rem beasts," if you want to know.

Decontamination is considered the least qualified of the nuclear "servitudes." With seniority (which does not mean permanent employment), workers can quit decontamination for assignment in other activities involving safety and radiation protection.

> In the beginning, when you start in nuclear, you start way down. You clean. We started in servitudes. That's cleaning, upkeep, all that. Then little by little you make yourself a little . . . how can I put this? You make yourself a little profession and then you're always there doing about the same thing. So I got myself into SRP, and it'll go as long as it goes.
>
> What does SRP mean?
>
> We do safety and radiation protection. We take care of everyone, all the companies. We're there to try to make things go as well as they can. We help everyone. (Denis, nuclear servitude worker, permanent contract with an outside company, 25)

Suiting

Anyone entering the controlled zone has to take off personal clothes and put on the suit required by the site they're going to (standard, "helmet," "Mururoa" [bubble suit], etc.). Suiting up is done under supervision from radiation protection agents and with their help. All persons who have been in the controlled zone also have to be checked on leaving, and unsuited, especially if the suits are contaminated, once again under the watchful eye of radiation protection agents. Standing in your underwear in front of a clothed person charged with checking your identity and whether you follow regulations puts one in an inferior and even humiliating position. Here nuclear servitude agents are invested with a certain power over other "outside" workers, not only because of their oversight function but also the associated task: dressing and undressing. But do they have the means to exercise this responsibility? Hermano's remarks leave room for doubt:

> Do you know what companies the operators come from?
>
> Yes. When we suit and unsuit them, they have to have an access permit, and the name of the company is always on the permit, along with the guy's name and their dose level to see if they're fit to go in or not. And we've got to check all that. (Hermano, nuclear servitude worker, fixed-term contract, outside company, 23)

In addition to the fact that skilled maintenance workers and servitude workers are employed by different companies, the social relations constructed in these situations create deep rifts between them, since one group is being monitored by the other in what are sometimes profoundly humiliating situations. Servitude workers have no power to sanction an employee from another company in case of rule breach. To clear their own name of responsibility for the breach, the only thing they can do is report it to EDF radiation protection services. This kind of practice presents an obstacle to the development of solidarity relations between these types of employees.

Sorting and Checking Radioactive Waste

Lastly, there is sorting and checking of radioactive waste, another task that Hermano is assigned to do. He tries to avoid high-dose worksites. But some tasks unavoidably involve exposure:

> We've got to check the bags of waste that go out. There are several sorts of waste: metallic, burnable (work suits, gloves, etc.), and non-burnable . . . that goes into containers; and then there are bags that've got over 200 millirems, bags that are less than 200 millirems, all that has to be checked, sorted. I mean, it's the machine (a programmable robot) that does it and it's put straight away into lead containers.
>
> And when you sort, are you wearing a helmet?
>
> No, the machine is hermetic. We've got our hands in it with a layer of lead like so in front for protection, and we sort. We have rubber gloves so as not to get contaminated but the radiation necessarily passes at the forearm level.
>
> Can't you protect your forearms?
>
> We'd have to work with lead gloves!
>
> And isn't it dangerous to get your arms irradiated?
>
> No. Hands and feet aren't dangerous.
>
> So why the gloves?
>
> They're for the contamination. You can stop the contamination, not the radiation—unless you put on tons of lead, or water.
>
> Water?
>
> Water is the best blocker. The reactor tanks are full of water. (Hermano, nuclear servitude worker, fixed-term contract, outside company, 23)

Hermano speaks of women being present in some plants: "the ones who do the cleaning, wash the floors" or "the work that men wouldn't want to do." They are usually of foreign origin, and there are never many.

Constructing Qualification: What Training, What Work Recognition, What Recognition for Know-How and Certified Skills, and What "Return on Experience"?

Nuclear servitude workers receive training only in PR and QSP. None of the servitude employees interviewed had any initial technical training and their education level never went beyond the CAP or BEP[27] vocational degrees. The incentive to get into this work was financial:

> How long have you been a DATR?
>
> This year will make four years. I used to be a charcuterer's apprentice. I

[27] Basic training levels in technical and/or vocational education.

got my CAP and worked a year in manual. But I wasn't making enough money so I quit.

It was a question of money?

Yes. And then working Saturdays, Sunday mornings, holidays—that didn't interest me. I had a pal who was already working in the plant and who got me into it. I took my DATR and I started a month later. They needed people. Since then, I've been on contracts. (Hermano, nuclear servitude worker, fixed-term contract, outside company, 23)

I'd been in sales for ten years. The company folded. We'd really tried to keep it going, though. I found myself unemployed. My mother-in-law knew a guy who was making good money. She suggested nuclear. I took the RP2 in 1990 and found myself at the head of a team immediately. I got a permanent contract right away.

When I got started in this I thought it was just to see what it was like. Then you start making money, you do a lot of hours—often 210 hours a month. Often you work seven days a week, weekends, holidays. You only do it for the money." (Patrick, nuclear servitude worker, permanent contract with an outside company, 35)

Jacques (nuclear servitude worker, temporary contracts, 33) was unemployed. To find a job he himself paid for the training that qualified him as an DATR:

The unemployment office doesn't pay for it for you unless you're on the RMI. [Revenu Minimum d'Insertion, basic French welfare program]

Training costs, at first paid by the principal EDF, were transferred to the outside companies or temp agencies. Given the current labor market competition for employment in nuclear servitude, the costs now fall to the workers themselves.

Not having any initial specialized training is also noted by servitude workers who have reached a certain level of seniority as a factor that devalues the skills they have acquired:

In our company, they've hired young guys who don't even know what it means to work in a nuclear plant. They hired them, they said "Here, it's free, we'll have you take the DATR, the QSP and then everything else." They passed them through, then they arrive on the site and they don't even know what a wrench is. I'm exaggerating a little, but you know what I mean. "Go get a torque wrench!" you tell them. Everyone knows what that it is, so do I. "What's that?" he says. And someone like that's working in nuclear! Those guys should be left at home to make pastry. They've got CAPs in pastry-making, bread-baking, charcuterie, catering, things like that. They got into nuclear because of well positioned EDF agents.

Quality guys like us, for example, we charge 50F an hour [slightly over 7.50€, in 1993]. They won't pay us that much. But if a little guy who was

> a butcher-charcuterer wants to get into it, they'll give him 37F [5.60€]. And they'll make money on the deal, let me tell you. That's how it's done. (Denis, nuclear servitude worker, outside company, 25)[28]

Denis and his colleagues also experienced what he described previously—without any initial training. What they are asking for is real recognition of the skills and know-how they use in their work. Those skills and know-how are extremely diversified, as are the tasks that their job involves. They have to understand in detail an extremely complex organization; they have to be able to get their bearings in it, be familiar with key spots in the plant, key tools and instruments, procedures for resolving the problems they encounter: how to deal with someone who has been contaminated, what to do when you can't manage to fasten a span of lead sheeting, what to do with contaminated objects—rags, tools—left behind after a maintenance operation. Confronted with these chance events, they are thrown back on themselves for finding ways of handling them that will bring the site back up to safety regulations.

PR and QSP training seem abstract to them, accessible to those with a certain level of education but unrelated to the practical work the job requires. They underscore the discrepancies between the quality doctrine they are taught and practice on the ground, where quality stands in contradiction with time constraints for getting the job done on schedule:

> That guy used to work in comic books.
>
> But he passed all the tests and training programs?
>
> He passed because they're eggheads . . . three days of training! But the tests are useless and the training programs are too because afterwards, when you're on the job, it doesn't happen like that. Even when they tell you, "You've got to do quality, guys, you've got to do a lot of things, you've got to keep straight." And then you get on the job and they say, "Listen, no, do it like this and we won't say anything, that'll be fine." That's why the training is strictly useless. Especially as far as quality's concerned. When you find yourself actually on the job, nobody talks about quality anymore. It's absolutely useless.
>
> That seems kind of contradictory, doesn't it?
>
> Yes, it's in contradiction because they tell us: EDF is a big company, we want people to do quality work, we want 100% good work, and then they prefer to hire the cheapest companies, which means there'll be less personnel. If the company wants to make a little money they're going to put on less personnel, the job done won't be as good. . . .
>
> EDF policy is money, it's no longer "quality," like they say now, it's the cheapest company that gets the job. And it'll be like that until the day there's a big problem, and afterwards they'll see that they should've taken

[28] Excerpted from Catherine Pozzo di Borgo's film, *Arrêt de tranche.*

> the ones who charge the most, because their employees work better. (Denis, nuclear servitude worker, outside company, 25)[29]

Another worker goes further, denouncing certain "clientelist" practices used by EDF agents in selecting subcontractor companies and what they call the instituting of mediocrity:

> For EDF it's not the quality of the work that counts. The EDF people say: "You've got a nice face, you pay for my dinner." There are bribes . . . and not only in soccer! In the purchasing department, they only reason in terms of money. At the general services department, you can hear them say: "Listen, thus and such company bought me a hunting day in Sologne." At EDF they're used to mediocrity, and they do everything they can to keep things that way.
>
> I don't have anything against my company, but you don't know what they think of you. I have an excellent project manager, but for the human side of things he's dreadful.
>
> We could do really fine work if we had the conditions, but EDF wants to stick with mediocrity. They've got to keep expenses down. So we run around, we don't have the resources. If EDF stopped trying to economize-really nickel and dime stuff—if they did a contract with a sufficient number of persons. . . . It's a question of poor organization. And we're stupid enough to say, "We've got to succeed." Professional conscientiousness does exist.
>
> I like to do my job as well as possible, but these are impossible conditions for that. Still, they know they can count on me. (Patrick, permanent-contract nuclear servitude worker, 35)

On the one hand, these workers are conscious of having acquired the know-how required for producing the quality work EDF demands, as Hermano explains:

> I've done all the jobs. After four years, I'm starting to know things pretty well. I've done assistance, radiation protection, checking waste bags. . . .

In positive perceptions of the job there is also a sociability dimension, even though it is constructed not in a structured permanent and stable work collective but rather in the course of moving from site to site:

> Do you like this job?
>
> I like it fine.
>
> What interests you in it?
>
> You're always together with chums. I go from site to site now, I know the people on them. (Hermano, fixed-term contract nuclear servitude worker, outside company, 23)

[29] See footnote number 28.

Hermano also says that the company he works for is known for the quality of its work:

> It's not just to give them a nice compliment! It's a company like any other, but you really do hear a lot of talk about them on the worksites. We're known as the spirits of the hearth.

On the other hand, there is little or no work recognition, as other servitude workers underline:

> There's no job recognition. There's only one person at EDF who told me I'd done good work. Otherwise, they're constantly getting on our backs for little things. They don't see the work we do. (Patrick, permanent-contract nuclear servitude worker, 35)

They also feel the lassitude of "routine," the fact that they are not acquiring new knowledge. Hermano worked for the same company on individual job contracts from 1988 to 1992. "They're always the same worksites," he says, adding:

> From the start I've been on single contracts. I'd like that to change.
>
> How long are the contracts?
>
> The length of the job. But since there are renewable jobs, we work ten months a year.
>
> And the rest of the time?
>
> I'm unemployed. I get the basic wage. That's 5600 F [approximately 850€ a month; in 1993]. When I work, with the bonuses and all that, I make 11,000 net [approximately 1700€]. But that has to cover travel too.

Seasonalizing and shorter shutdowns enable the outside companies to reduce travel and housing costs dramatically. This decrease of course works to lower shutdown cost for EDF. The resulting unemployment for the workers, on the other hand, corresponds to a major fall in income. The employer can thereby diminish payroll in periods of low activity, which is supposed to be impossible in the framework of a special status state company like EDF. To this must be added all the resources for "flexibilizing" temporary workers or leased labor. To my knowledge, none of the companies employing the workers I interviewed was checked by labor inspection services for ongoing use of temporary workers (a legislation violation[30]) or inter-company leasing of workers.

[30] Article L 124-2 of the *Code du Travail* specifies: "Temporary work contracts may not have for purpose or effect the filling of a job position linked to user company's normal, permanent activity." Many DATR workers are "intermittent" (Doniol-Shaw et al., 1995)—i.e., they are offered a succession of fixed-term contracts by the companies that employ them—while occupying positions fully linked to the "normal, permanent activity" of those companies.

In Hermano's company in 1992 there were approximately 150 fixed-term contract employees for 150 permanent-contract employees. But "when there are a lot of worksites, there can be up to 500 temporaries." Furthermore, subcontracted companies regularly lease other companies' employees to increase their own workforce when an emergency demand from EDF requires more workers than the company can provide.

When Hermano was interviewed for the first time in 1992, he was satisfied with his situation:

> I prefer working on single contracts. That way, when the day comes that the work is no good anymore, I can change companies. Since there are lots of companies who do the same work . . . I don't want to be dependent on one company. I already don't like being told what to do! When I've had enough, I just ask the boss to give me back my working papers.

But annual work time has fallen continually over the last two years and unemployment payments run out. When I interviewed him again in 1994, he had given up working in the nuclear industry for his old job of charcuterer in a supermarket:

> Why did you decide to quit nuclear plants?
>
> Because in the long term I thought that it was going to start affecting my health. And I was sick and tired of moving from site to site—and stopping for six months. Because you worked six, seven months a year and the rest of the time—unemployment. That was the big drawback in the job. The unemployment payments regressed fast. And the last year, I wasn't eligible for them because I'd been on them three years and I was seasonal.[31]

Patrick and Denis are on fixed-term contracts. They live in fear of their company's losing market share, which would threaten their jobs. They have therefore internalized the idea that the purpose of their work is to satisfy EDF demands, in a spirit of competition with other companies, the aim being to position themselves at the head of EDF's preferred-company list:

> As long as the customer's happy, as they say—and the customer is EDF—as long as the customer's happy with you, they call you back. They try to have the same personnel, like that they know what we can do. . . . But this way of doing things is disappearing. If the others are cheaper than us, even if they don't know the company, they take them. I'd been going to Nogent for four years as an SRP [Servitudes et RadioProtection]. Well as

[31] In her 1997 thesis, Penser le rapport au travail pour comprendre le chômage [Understanding unemployment in terms of employers' and employees' relations to work], Nathalie Frigul brings to light the processes of losing unemployment compensation entitlement for persons working intermittently (cf. Frigul, 1997, ch. 7: "Disparition ou altération des droits liés au travail," § 7.1.1: "Les effets du durcissement de l'assurance-chômage").

> of October of this year we aren't going there anymore because we were too expensive compared to another company—regardless of the fact that the EDF head in charge of SRP was not at all in favor of hiring another company. But it was purchasing that decided, and purchasing told us, "We take the cheapest company." So we aren't going to Nogent this year. (Denis, nuclear servitude worker, outside company, 25)

This comment brings out the contradictions between the different EDF departments in the area of maintenance and in the criteria used to select subcontract companies. For the purchasing department, the point is to reduce costs at all costs. In contrast, the EDF agent in charge of supervising tasks performed by outside companies would prefer the guarantee associated with prior experience.

Ultimately, it is willingness to move around that tends to determine qualification (and wage). Denis has been going around in a trailer with his family for eight years; Patrick left his family four years ago to find work. When he asked to be assigned for a while to a plant near where his family lives, he was retrograded on his company's qualification grid:

> Currently I'm a Level 2 supervisor. I asked to stay at S. [*nuclear plant*] to be with my family a little. They said ok, but they took me down to a lower qualification level.

In nuclear maintenance, "servitude" is a "specialty" that is constantly needed during shutdowns. Can work volume really fluctuate given that, as nuclear plants age, contamination and radiation levels increase, making not only decontamination but also all other forms of maintenance assistance and operations absolutely necessary? In this sector, despite a permanent need for certain tasks, there is heavy use of temporary workers—a practice that stands in contradiction with the *Code du Travail.*[32] We are naturally led to question the motives for this practice.[33]

CONCLUSION

We are now in a position to identify certain characteristics of the relations obtaining between the way work is organized in the nuclear industry, the way risk is managed, and the type of consideration for workers' health—relations that are relevant for all subcontracted company maintenance workers.

Despite diversity of companies, activities, and statuses, "outside" DATR workers in this study attested to a shared reality: the world of maintenance work in

[32] In the STED Survey population (Doniol-Shaw et al., 1995), the proportion of temporary-contract employees reached almost 30% of personnel working in nuclear servitude. As we know, these positions involve extremely high exposure to ionizing radiation.

[33] See n. 26.

the nuclear industry, the division between EDF agents and "outside" workers; also the procedures, rules, constraints particular to this world, the most significant constraint being exposure to ionizing radiation. This seems the structuring factor for both work organization and the individual activity of each operator. It is a reality characterized by multiple forms of labor division—this is the focus of chapter 2.

Another dimension involves the experience and know-how of the workers interviewed. Given the details of the tasks they perform and the practices applied, clearly the standard simplistic equation between low job stability and low level of skill and professionalism cannot be justified here. All examples cited by workers suggest the opposite. In valve and pipe maintenance, inspection work, and nuclear servitude, "outside" workers interviewed express sharp awareness of the problems and dangers involved in the jobs they have to perform, as well as a desire to have the high quality work they do recognized for what it is. What they run up against is denial of recognition from an anonymous subcontracted labor organization system that eliminates all negotiation space between the subcontracting power that buys the labor and defines the work and the "outside" workers who perform it. This system allows EDF to rationalize (and ration) maintenance activity more every day, without having to negotiate the decline in the quality of hiring, work, and income conditions for "outside workers." Meanwhile, the workers are caught in the contradiction imposed on them between training that emphasizes a discourse requiring individual accountability for quality, safety, and security from "outside" workers, and companies' selection practices, which are based exclusively on the criteria of cost and, for some, clientelism, a way of proceeding that dismisses workers' experience and the real quality of their work.

This contradiction is echoed in the main characteristic of subcontracting practice as brought out by Marie-Laure Morin; namely, the customer-supplier type relation of domination, which pushes out the wage-earning-and-paying relation characteristic of employer-employee relations (Morin, 1994). Between EDF and "outside" workers there is no work contract with associated guarantees, but rather customer-supplier relations, mediated by the employers' market (or markets, in the case of sub-subcontracting). In the social relations of work that exist between EDF and outside workers, there is therefore no space for negotiation, despite the fact that labor regulations have institutionalized such a space; there is no space in which nuclear maintenance workers can express their views on doing maintenance work in nuclear plants, on the conditions of that work, on the way it's organized. Meanwhile, EDF managers can make any decision they wish. This is demonstrated by the social division of labor as related to radiation dose (the focus of chapter 2), and the intense "flexibilizing" of work (chapter 3). It will be seen that the impossibility of negotiating the conditions under which nuclear maintenance work is performed can adversely affect the quality of the work done.

CHAPTER 2

The Social Division of Labor and Radiation Dose

The social and technical division of labor, long an integral part of economic choices and industrial strategies, has always gone together with inegalitarian distribution of risk, due to its very founding principle, i.e., the division between idea and prescription on the one hand, realization on the other. This general approach to the division of industrial labor was continuously maintained, though modified by the development of the production system, labor relations, and the social welfare system. The labor relations that structure work organization lead us to distinguish several ways of dividing labor: socio-cultural, hierarchical, and technical. To these G. Mendel adds what he calls the "institutional division of labor," which designates differentiation of activities and jobs within an "institution" (a company or any establishment structured by a particular type of work organization). What characterizes the institutional division of labor is that "jobs are not understood to be practiced by separate individuals in the society, nor is each job performed for itself. They can only take on full meaning . . . in relation to . . . the institutional act as a whole" (Mendel, 1997).

In nuclear industry maintenance work, the institutional act refers both to organizing maintenance tasks and applying the set of safety measures recommended by the 1984 "quality" ruling (see Part II). This regulation stipulates that the work of executing tasks and the work of inspecting them must be performed by "different" persons. Lastly, division of labor in the nuclear industry is determined by management of exposure to ionizing radiation.

The labor relations established between different categories of workers in maintenance work are complex due to the extreme heterogeneity of this work community. There are workers with state civil servant status, others without that status, permanent and temporary workers, skilled and unskilled. The social division of radiation exposure puts some workers in a situation not only of "serving" others but also of being more likely than any others to be exposed to the radiation and contamination that are part of maintenance activities as a whole.

SOCIAL AND LABOR RELATIONS

Nuclear plant maintenance presupposes human intervention in extremely dangerous places. The division of labor therefore takes on a very particular character. What allows for performing the tasks necessary for ensuring safety is not only social, cultural, hierarchical, technical and institutional division of labor but also and perhaps first and foremost a subdividing of the radiation dose. First, there is massive externalization of radiation and contamination risk away from EDF agents toward "outside" workers in accordance with the classic technical division between idea (in this case preparation and inspection) and execution. Among outside workers themselves, another way of dividing labor appears wherein operations that are important for safety are distinguished from tasks performed earlier to make those operations less costly in radiation. Those tasks—chronologically first—involve high exposure to ionizing radiation.

This distribution of exposure is made socially legitimate and acceptable through strict conditions meant to ensure that regulation levels of exposure are not exceeded. That is done through a complex procedure of arranging job assignments in terms of likely "dose"; only "outside" workers can explain in detail how this works. The maneuvering room they have for managing the contradiction between health and employment and negotiating their own "relation to the dose" are determined almost exclusively by the place they occupy in the organization of labor.

Social and Labor Relations—The social and technical division of maintenance work in the nuclear industry is constructed through a dynamic of labor relations between not only supervisors and operators but also principal's agents and "outside" workers, as well as among outside workers, all with regard to production goals, principles of work organization and the types of power that govern work relations in nuclear industry maintenance work.

Between Safety and Managerial Rationality: Hierarchical Relations and Subcontract Relations

An account by valve and pipe mechanics brings to light the complex configuration of hierarchical relations that determine subcontracting relations and the division of tasks between outside operators and permanent principal agents. The operator speaks of relations between "outside" workers and constructors (namely Alsthom and Framatome) and certain EDF national technical departments (designed as "UTO" and "GDL"); between workers and the EDF departments in charge of section shutdowns (specifications and planning), preparation, inspection), radiation protection and cost control; and finally between workers and project managers at those workers' own companies.

Contradictions arise in work activity related to the respective aims and motives of the various employers and the means and maneuvering room that employees have for reaching those aims. I am referring here to analyses by Yves Clot, Jean-Yves Rochex, and Yves Schwartz (1990) showing the importance of identifying such "motives" in order to understand how employees handle the various levels of demands and constraints they encounter in their work.

The following example from a mechanic illustrates these hierarchical relations and the particular division of labor they produce:

> It was the first "Case 1" modification I did on a nuclear site in the controlled zone with a constructor. We were taking off an RRA valve [valve on a shut-down cooling circuit]. As we were taking it off, we happened to notice there was a 100-mm scratch on the actuating lever arm—which means there' s a risk of leakage. So the constructor decided to replace the lever arm. EDF said no problem. But it was a valve classified "mine." We go to the equipment stockroom to see if they've got lever arms. They had five. But there's a quality file on them. I go get a lever arm and I bring it back. I see my foreman and ask him if we can put it on. He asks me if I've got the file. I answer that I was told it was the preparer (EDF) who's supposed to give us the file. The person in charge of quality at the constructor's answers "No problem, put on the lever arm." I do it. We sign for the checkpoints. We put the valve cap back on. Once it's back on—because the valve had to be put back on that day (the operation takes three hours, no more)—I go back to the office and tell them the valve cap is back on. At that point the foreman says: "I think we're going to take it back off." "Why?" "Because they lost the quality papers for the lever arm." That meant that for the constructor the arm was no longer regulation. They'd lost the papers.
>
> The fact is, I was supposed to close the valve. Which I'd done. So that day, after I turned in the job papers, they'd turned the water back on right away. So we couldn't go back and operate on the valve. So they wanted us to use an ice plug. They wanted to turn on the nitrogen captors and freeze the water coming in and going out. But no one had ever done that. And how do you know there won't be leakage and all? All that for a piece of paper! An important one, but still!
>
> The constructor's quality manager has been in nuclear for 23 years. Very competent in his work. And he managed to get EDF to admit they were in the wrong and that we were in the wrong too. They wanted to fault us. But having to wait for the paper. . . . On the one hand you've got a guy who's saying "Got to have the valve finished by this evening" and on the other "Got to wait for the paper," and until you get the paper you can't work. That blocks five people, you see. So they made the decision. We'd gotten EDF's accord to put it back on in any case, so we were covered. But a verbal agreement is easy. I can give you all the agreements you want; as soon as there's a problem they can say "I didn't give you anything."
>
> So you got the agreement orally?
>
> We'd gotten it. The guy from EDF who was overseeing us took responsibility. He'd said "Yeah, yeah, we'll take care of it. You can come get your

> paper this afternoon." That meant the other guy gave the okay to the constructor who gave us the okay, so we put the valve back on.
>
> We left the valve just as it was and a decision was made to write an irregularity report. The fact is, the valve arm's fine, there's nothing the matter with it, it cost a fortune . . . you could buy yourself a nice car for that! But the decision was made that during the next unit shutdown the valve lever arm will be replaced. Even if there's nothing wrong with it. That arm would've functioned for a year, and they're going to throw it in the garbage because the papers are missing. And there're lots of errors like that. That was the first time I did "Case 1"—I felt pretty strange. I mean, we weren't in the wrong. We'd gotten the go-ahead. All the people above us had accepted their responsibilities. But what I mean is, the strong point in our favor was that we were working with the constructor in Case 1 modification. What they've got going for them is the fact that they're constructors. Compared to us, they're in better shape than if we had made the same mistake, in Case 2. We don't have the goods, we don't have a quality department. And he's the constructor, so he can always say, "The valve will work even with that arm." They need a guarantee at the equipment level. Was the arm made out of the same material, that's the problem. Was it signed for? Was it regulation?
>
> Who's going to pay for the new arm that will be put in during the next unit shutdown?
>
> EDF. The constructor decides and EDF has to do it. If EDF doesn't want to, EDF is accountable for potential future problems. It's no small piece. It's 80 centimeters in diameter and 2 meters high". (Guy, mechanic, permanent contract 35)

This example shows the division of labor and the contradictions it creates in work activity due to the conflict between the various "production motives" (Clot et al., 1990), here those of distinct companies linked by a subcontract relationship that creates power relations among company employees as a function of their respective positions as either principal's agents or subcontracted company employees.

For the constructor, the priority was not only to guarantee the technical safety of nuclear installations but also to have a written document proving that the material used complies with quality guarantee specifications. This explains the importance of the document attesting to quality: in a way it represents the constructor's quality insurance. In the case just presented, if ever there are problems, the worker will have to be able to produce guarantees of valve lever arm reliability. If the documents (some of which are signed by nineteen safety authorities) stipulating that the arm meets technical specifications cannot be produced, the arm cannot be left to function in the plant. Still, the problem of missing proof (which, from the perspective of safety, should have led to immediate removal of the piece in question) was resolved by a compromise that consisted in leaving the arm in place until the next unit shutdown. What functioned as the basis for this compromise was the experience of the constructor's quality manager—someone "solid," as the interviewed employee stressed. Nonetheless, the compromise seems in contradiction with the "safety" doctrine. In safety terms, what does it mean to plan to change the arm a year later? Either the arm is the right spare part, and in that case it can stay in for the duration specified for that type of arm, or else it is not the right spare part, in which case it poses immediate dangers. In fact, this type of problem is anticipated and planned for in the quality decree of August 10, 1984, and should have led to drafting an anomaly report. But the constructors know that EDF is reluctant to proceed this way because it would raise questions from DRIRE inspectors. Clearly what is operative and indeed dominant here is not so much safety logic as the EDF managerial imperative as internalized in this case by the EDF decision-maker.

From the perspective of EDF administrative management as ensured by the company representative, we can hypothesize that the fact of not being able to find the necessary document is a foreseeable event during unit shutdowns given the extreme diversity of operations, maintenance, and other types, the numerous related papers, and the relatively high number of persons present on the site at the same time for different reasons and with different purposes. The event confronted the EDF agent in charge of supervising the operation with a choice between two strategies: suspending the maintenance operation until the paper was found, which would have meant delaying and re-planning all operations of the other companies that were scheduled to follow this one, or giving the go-ahead and hoping to locate the document as the task was being performed.

The choice had to be made fast. Given the managerial imperative that the EDF agent was assigned, it seemed a lesser risk to give the go-ahead than to suspend the "flow" of consecutive operations. Moreover, the water was restarted as soon as the valve mechanic finished, thus making the arm-installing operation irreversible, unless there were a real emergency interruption, which would have involved using resources for a procedure that no one was familiar with: making an ice plug to stop the water.

The outside worker who changed the valve lever arm found himself up against two contradictory injunctions. His foreman invoked the safety regulation—being in possession of the required document—but the EDF agent (preparer) offered to guarantee the availability of the document and the constructor's employee ordered the mechanic to install the arm. He did it, but he was reprimanded afterwards, and only then did he become aware of the risk he had run for himself and his company: if the EDF agent hadn't taken responsibility for his decision, he could contest the oral agreement he had given. The subcontracted company has neither the constructor's power nor its experience to legitimate the operation and its consequences, nor the power to decide what will happen to the controversial valve arm. Because the matter is "Case 1" the constructor alone is responsible and the subcontracted company employee has very little maneuvering room.

The fact is that the way this operation was handled infringes on a fundamental general safety rule: obtaining the necessary documents before taking any action. The fact that this rule contradicts the obligation to produce results within a strict time-frame—a major imperative for reaching the management goal—is not taken into account by the principals (the constructor and the owner, EDF). EDF is not accountable to anyone for the way it is organized—here, the way it managed papers pertaining to the quality of stockroom equipment. The only way the DRIRE could have intervened to monitor the situation and audit EDF quality department on matters of organization would have been to file an anomaly report.

On his side, the valve mechanic knew that any delay in executing the task at hand would have penalized his company financially either in the short or long term because it might have difficulty later getting new contracts with EDF. This would also threaten *his* job. He thus handled the chance occurrence by giving priority to the time constraint while of course having no power to oppose the constructor or the owner on the breach of safety regulations.

The radiation protection constraint is not at issue here, but in other situations it represents a third pole of contradiction, as we saw in chapter 1 with the example of an operation during which the usual procedures for installing a SG plug were disrupted by another chance occurrence: screws getting stuck, making it impossible to install the plug.

In the examples of the SG plug and the valve lever arm, the issue was how to regulate a problem collectively when the time spent discussing it could have an impact on getting it done. It should be noted that it fell to the outside worker to find a strategy for handling the chance occurrence without breaking any safety or radiation protection rules or casting doubt on his own company's productivity and work organization.

Coordinating Operations

Every specialized activity such as pipe and valve mechanics is integrated into a tight fabric of other activities in the framework of a division of labor among employees of the various outside companies. The tasks to be performed are often interdependent and hierarchically ordered in social terms: there are maintenance activities strictly speaking and service activities.

Among employees technically trained in mechanics, electricity, welding, and inspection, the different activities are coordinated in relation to EDF's work schedule. Employees of different companies rub shoulders and occasionally help each other out: lend tools, set up the next set of activites if they are operating on the same site on different system components. But the intense time pressure means that many are operating at the same time in the same places, and this also creates tension among employees from different companies, tension heightened by the radiation exposure constraint:

> You get to a place and there are a total of six different companies working there. Any more than that and they cut the number. . . . It used to be you'd get to a site and you had it to yourselves for the duration of the job. Now there are several of us. It happens that I find myself working next to this guy on the same type of valve. And when it comes to dose discharge, if he's doing volumetric, he's going to send it all over to my site.[1] Or if he's doing fixed, he's going to walk around and send it over to my site and they're going to ask me why there's fixed[2] on our side. That becomes a real drag . . . you're contaminated—they don't like that. Then you've got to tell them where it happened and all that. . . . It's like with your [driving] license when they take off "quality" points. . . . The thing is, there are too many of us working on the same site. Six of us on the same site, we're all on top of each other. You lay a piece of vinyl on the floor and the other guy lays one down right next to you—we can't mark out our site! When we mark out, we take it wide, we need room. We space out three valves to the side. Like that, there's no problem. (Fabrice, permanent contract mechanic with an outside company, 24)

> That way, if there's an accident, a guy who gets hurt on our site, we're not responsible. (Eric, permanent-contract mechanic, 22)

> But now we all have to work at the same time. It's no good. We say to each other: "Hey, lend me your wrench, do this or that for me." We used to get there, do the service, do a pipe, when we left we'd do a site withdrawal, everything was clean for inspection, there were no problems and it worked well. It's true that it lasted longer. Now it's a question of dough. (Fabrice)

Outside workers especially dislike some situations because they involve anonymous irresponsible behavior by other employees, but it should be remembered that they don't know the conditions the others are working under.

> The problem is that no one takes responsibility for this [waste management]. For example, you're on the "four aces" (primary circuit valves just below the center section of the primary piping)—I do a lot of four aces. So we're getting ready, we're getting the waste bags ready too, but you're not the only one working in the reactor building. A person goes by who's got rags to put down and the first waste bag, so he just plunks them down—he doesn't look. Sometimes even before the operation the bag is already full—we haven't even started to work yet. All because there are people who can't be bothered to go get a bag.
>
> Things like that can't help but happen on sites where you've got thousands of people. You put a piece down here, someone goes by, the piece was in the way a little because there's a plug or something near it so you find your

[1] Amount of available space varies by nuclear plant site. Older sites are more spacious than newer ones, where the concern was to use concrete sparingly.

[2] "Fixed" refers to a type of contaminated dust that sticks to the floor.

> piece two meters away. Everyone's got to work and be able to put their material down somewhere. If the other guy's on the RCV circuit [circuit for regulating primary water volume), it's very active and he can send contamination onto my site that can affect a less active circuit.
>
> Tools get unplugged because someone else needs a plug, even when it's marked "Don't touch." You're in a Muru suit when all of a sudden there's no air. You go see what's up and find that someone's unplugged the electrical source which works the assisted respiration system—because they needed the plug. These people are really not conscientious. They're even a little crazy! (Guy and Didier, permanent-contract mechanics, 35 and 26)

Operators often mention the problem of not having enough electric plugs, but this seems never to have been corrected by EDF—a concrete example of the effects of not being able to negotiate operating conditions.

Moreover, each company is required to follow EDF regulations and hierarchical functioning principles in connection with its demands as principal:

> We have a preliminary site meeting with our foreman, all the guys plus the guys from EDF. We do the preliminaries, we talk about the job, the time frame, the work we've got to do. The foreman tells us what site we go on, and if we've got problems we go see him. He's the one who clocks us in and out. . . . The work supervisor is in charge of himself and the operators. He's got lots of stuff to do. He does site safety. If it's in the [controlled] zone he does all the radiation protection before the work begins. . . . In theory the work supervisor has to recheck everything all the time. For his own safety. We go get a specs sheet and the first thing he's got to do is check the sheet, see if it's right. He's got to check all the key points listed on the sheet. They demand so much from the work supervisor. He's got to work, take care of all the paperwork, it's impossible . . . They've even suggested that the work supervisor shouldn't work anymore because he's got so many documents to take care of. There's more paper than work! . . .There's stuff on paper, and then there's the work. If we applied it all to the letter we'd never get done. The unit would be out six months instead of three weeks!
>
> Who checks the work once it's done?
>
> There's EDF follow-up. We've got an EDF agent who follows us. But all through the operation there are phases. There's the opening of the site, where they give you either a checkpoint or a supervision point. With a checkpoint, it's mandatory, the EDF guy has to come see. For a supervision point, it's optional, we tell him by telephone and he says okay. He doesn't have to come. . . . For a shutdown, he has to come sign. He comes to evaluate and sign. If it's clean it's ok and we can put it back on. If there's damage the matter goes higher up. They've got to see if they're going to change it, because it's extremely expensive. Recently I did a valve that cost 15 000€ to do. That sometimes gets to them.
>
> EDF are the ones who decide?
>
> Yes, we don't make any decisions in these matters. After we're called in, usually to take the valve apart, for example, they come see if it's taken

> apart right, but that's optional. And they don't usually come into the zone. They don't want to get dressed for it. We call them and they say okay. . . . The EDF guy isn't following just one team or company. Sometimes he follows three, four companies. So sometimes he doesn't have time. But for a checkpoint he has to come. On checkpoints we can't continue until he's signed. Sometimes you lose a day on it. (Eric, permanent-contract mechanic, 22)

EDF agents are vested with final authority on all job execution phases. This is an absolute power relation: any and all discussion or negotiation between "outside" workers and EDF agents is up to the absolute will and discretion of EDF.

Labor Relations between "Outside" Workers and EDF Agents

Relations between "outside" workers and EDF agents are structured by subordination. The workers' actions are of course preceded by prep work and, more importantly, followed by EDF inspection. According to the quality decree of August 10, 1984, that inspection activity must be distinct from the non-destructive inspection of material components described in chapter 1. Valve and pipe mechanics complain that inspection of their work never leads to any real recognition—except if EDF has to disclaim responsibility in case of a problem that arises afterwards.

> Really, we workers, I don't know, but really we should at least be kept informed. It's incredible that we're not even told sometimes whether the job went well. We move on to another job and we're not told if the first one was done correctly or not. Then you get a letter five months later saying you screwed up on this or that valve. We should be told at the end of the job if everything went well or not, if we did right. We're always informed a year later. (Jean-Luc, permanent-contract pipe and valve mechanic, 25)

In pipes and valves, where EDF supervises throughout the operation, the results are generally noted on site, in the plant. For non-destructive inspection activity, however, the results are transmitted to one of EDF's specialized national services or the constructor that requested the work. Consequently, relations between "outside workers" doing this job and EDF plant agents are limited to the administrative procedures associated with the operation. These procedures are perceived as extremely bureaucratic. The EDF agents in charge of ensuring that procedures go as they should are said to have no real familiarity with the concrete reality of the work involved. They and non-destructive inspection specialists constitute two hierarchically ordered "worlds" that cohabit during the space of a unit shutdown in the framework of a division of labor in which one group does the work that the other would refuse to do because of the conditions:

> There's another thing that gets to you: EDF personnel is above all administrative. I met a guy from the radiation protection service at M. who said, "I've got my little computer now, I'll never have to go into the reactor anymore." They don't go on site anymore. They spend all their time drumming on their computer keys, filling out permission slips for exiting and entering. That's probably the most serious problem. Those EDF agents have got to get back on the ground. It's fine to subcontract, but you've got to keep control of things. It's really two worlds—and you don't go through the same gate to get into them. There's an EDF canteen and a "companies" canteen. There's real separation. Plus it's clear that we're doing the work the EDF agents don't want to do. And they're the ones who sanction us! Often they get off at 4:30, company guys don't get off until 5:30. It's too unbalanced. Especially since the companies that work at EDF are not minor ones. They represent a major [personnel] mass in the factory. . . . More and more at EDF they try to make the companies accountable. I agree with that idea, but it's also a way of getting out of responsibility—for the work but also for safety. It's the company that has to ensure its own safety. We're getting to a point where the company is being left more and more on its own, where responsibility for any mistakes is assumed by the company. And I think it would be better if EDF were more on the ground instead of in the offices. (Marc, fixed-term contract inspector with an outside company, 33)

The EDF agents that are in contact with servitude employees work in the safety and radiation protection (SRP) department. When Hermano does decontamination he experiences them as "gendarmes":

> We have to go see the SRP agents every time we go in. As soon as you get to a site you have to go see them. First for a radiation protection test, to be sure you're familiar with the deal. Then you go through anthropogammametry to make sure you're not contaminated and afterward a visit to the doctor's. . . . SRP agents are there all the time. They check everything that involves radiation discharge and contamination. That's security. They're gendarmes. They can even be a real pain sometimes. There are jobs you could do in five minutes and they require us to do them in an hour, which means a higher dose. . . . For example, you've got a contaminated spot. We've already done samples to see if it's contaminated. Then we mark out the area. Then the foreman arrives. We tell him, "It's contaminated there." I take a bucket with a floorcloth and some decontaminator to clean up. But if there's an RP agent there, he's going to ask me to do the samples first, go get them checked, apply the first product, a second product—when we've got products that decontaminate just as well. We could do it in ten minutes. But there've got to be all these checks. When we've cleaned up already, they come checking behind us, and if its contaminated we do it again and again until it isn't contaminated anymore.
>
> Why do you think they prefer their method?

> Because they don't trust us. And it's not the same. They don't work, it's not them who do it. I'm sure that if they were the ones who mopped the floor they'd do the same as us. But they're just there, like gendarmes.

For Denis (permanent-contract nuclear servitude worker, 25), the EDF agents' way of following ouside workers is extremely bureaucratic in comparison to how things were done in the Belgian plant he used to work in:

> In Belgium, every company is carefully followed. I mean the guy supervising the company is on site, so things go much faster than here. Here there's EDF oversight on the company, and they can intervene at any moment if there's a problem, but you've got to call the guy, he's got to come, he's more likely to be at the office than on site. At the administration level, paperwork, there's definitely very close oversight. But I'm talking about site oversight, at the level of operations. We've got an EDF agent who's responsible for operations and he should be on site systematically to follow the company, even if he's only there to see that the work's being done right. But we don't usually see the oversight guy very often. Seldom, in fact. If there are papers to do, he's usually in the office rather than on site. Whereas in Belgium it's the opposite. Because in Belgium, the deal is productivity, more so than here because over there they're privatized. The shutdown has to go as fast as possible; that means taking as little time as possible but doing the job right. That's why we see most of the agents, the guys in charge, on the work site. They're there with the companies and they see if there's a problem. They're there right away to tell you, "We're going to do this, we're going to do that" without having to say: "We'll have wait, we'll see, we'll have to have a meeting, we'll see what should be done." Things go much faster. They make decisions right away.

This account brings out the difference between French and Belgian nuclear plants when it comes to regulation systems and responsibility-taking. As we saw in the example of changing the valve lever arm, after a delay due to hesitation about how to proceed, the constructor rather than the owner takes responsibility for the decision. In France, operators are affected by the time loss due to delayed decision-making. Moreover, the way the system functions allows the EDF owner to externalize its "safety" responsibility. In Belgium, on the other hand, according to Denis' testimony, the owner is present and assumes responsibility for necessary decisions.

Patrick (permanent-contract nuclear servitude worker, 35) underlines the gap between the quality demand that EDF agents are constantly reminding workers of and EDF immobilism with regard to anything other than the imperative of working fast, and the related fact that subcontracted companies are offered no resources to meet this demand:

> EDF's people in charge of the operation . . . there's not one capable of making a decision because they're afraid of taking responsibility for it. There's the

> fear of a nuclear disaster, but also, when you see how they work. . . . Forcing a nuclear component in order to go fast and spend less money. . . . When you do that you're opening the door to anything happening. They make you these "quality" plans, then on site they don't give you any resources. The reactor building has never been so dirty. And you've got to run all over the place. So we run. We don't have the resources.

Antagonisms and Social Hierarchy in Maintenance Jobs

A young valve and pipe mechanic is proud of having the "best" job of all the types of activity:

> Compared to the other guys you've got in the plant here, the lagging workers, the guys who put up the scaffolding, the cleaners—compared to them we're proud because the fact is we're the best. We've got the best job in the plant. (Eric, permanent-contract mechanic, 22)

There is discrimination between employees who perform technical operations and those who "serve" those operations. Pipe and valve mechanics do not have to clean contaminated surfaces or sort waste materials, and this limits the dose *they* receive. But in fact they consider these tasks secondary; they are "servitude" tasks. Decontamination and waste sorting are thus not socially valued activities—the case for all cleaning and waste-elimination tasks whether domestic or industrial (Bretin, 1994). Pipe and valve mechanics think of these services as "natural" and think that they "naturally" have to be performed before their own operations. They make no reference to who takes does them or what it means to "take radiation dose for the others"—as one servitude employee put it—or sort waste:

> The waste team takes care of all the waste. On our side we have to get bags for the oily rags, for papers . . . So we put it all in a bag and then we go see the guys at the place they stock it all and we explain what it is. In fact, it depends on the sites. There are sites where they're paid to do it and others where it's up to us to do it.
>
> Yes, but they re-sort after you. The guys over there, they recheck every little object for dose. If they've got one that has a rem on the bag when it's filings, they re-sort to find that part. That way they can separate out the highly contaminated objects afterwards. Those things don't go into the same containers. (Guy and Didier, permanent-contract mechanics, 35 and 26)

In nuclear servitude companies, as we saw, some employees monitor the outside workers of other companies to see that they're following radiation protection regulations. Pipe and valve mechanics express mistrust of those who exercise that function. While EDF agents' oversight power in matters of radiation protection, safety, and work evaluation is recognized as legitimate, the "SRP companies"—designated by the nickname "Schtroumpfs" [Smurfs]—are feared and sometimes scorned for their work of "policing" other outside company employees to make sure they are following radiation protection rules. A pipe and valve mechanic who underwent what he considers a poorly done, excessive check by an outside company SRP employee that resulted in an "accusation of contamination" expresses doubts about that agent's DATR qualifications—"You wonder where he took his PR2!"—especially since the contamination incurred a reprimand for the employee and his company:

> Often, if you ever get contaminated, what happens more and more is that the company or the person who's contaminated automatically gets blamed. Sometimes you open a circuit that shouldn't be contaminated and you find that in fact, it is. . . . If ever a guy on my team gets contaminated, EDF's going to hold me responsible. If ever I open a circuit without paying attention or if I've got people who aren't well trained, it can go as far as being prohibited to enter the zone. (Jean-Louis, permanent-contract mechanic, 37)

This technician must often supervise sizeable teams of temporary workers with all the difficulties this involves for ensuring that radiation protection rules are followed by workers whose experience level he doesn't know.

For their part, servitude workers do not appreciate the conflicts their inspection operations cause:

> Sometimes there are little tangles with the other companies. We're there to suit them, show them how to do it. Sometimes they don't follow the rules, so we have to call the RP agents [EDF]. Because there are some who just won't listen— on pretext of being bosses at Framatome. We have to check them all. It's true that sometimes there's some shouting with the guys. . . . Some of them don't want to put on their harness because it gets in the way. The plastic suits can be pain when you've got to go through the little holes. Some put on a mask when the fact is you're not allowed to, it's the ventilated suit that's compulsory. There are RP agents there, but not always. At night there's no one. And we're not the ones who're going to act the police. I put myself in the guys' place too. (Hermano, fixed-term contract nuclear servitude worker, 23)

Servitude workers feel they've been put in a false position because they have been given a task that, in the very logic of the division of labor on these sites, should fall to EDF agents. In this way there's always someone who can be blamed other than the EDF agent.

The competitive relation in which service companies are engaged with each other to get EDF business can transform the relation among the various companies into internecine worker warfare:

> We're there with competing companies and it's major squabbling. For example, there's a guy from another company who's laughing at us because we've got problems at work. I've got two rugby men. He'd better keep his

> comments to himself or he'll get his butt kicked. There's big competition! (Patrick, permanent-contract nuclear servitude worker, 35)

This remark illustrates not so much the breaking up of work collectives as the profound antagonisms that get constructed among workers on the same site. These antagonisms arise out of subcontracting relations where owner's domination is founded on its putting subcontracted companies and their employees into constant competitive relations with each other. Among the workers, it's "war," and what's at stake is keeping their jobs by not doing anything that would prevent the companies they work for from getting new contract offers from EDF.

These antagonisms affect power relation protagonists on the work sites. Power relations are no longer between workers and their superiors but among the workers themselves, anxious to win "favor" with "EDF, the customer." Living under permanent threat of their company losing contract offers, which would threaten their jobs, they internalize the imperative of satisfying EDF demands as the purpose of their work, in a spirit of competition with and against workers from other companies, so as to come out on top of EDF's "preferred companies" list.

All these forms of division of labor and the relations of subordination associated with them are accentuated when the workers in question are temporary. Temp workers' status is particularly fragile and they have the least maneuvering room of all workers. This is clear with regard to pipe and valve mechanics on fixed-term contracts working on low-technicity, high-radiation exposure tasks, especially nuclear servitude workers. The term used to designate this service activity is both highly symbolic and underlines the immense social distance separating "technicians" from "slaves" (as one of their number described himself). As explained, the second group prepare the ground for the first.

Nonetheless, all persons working for a service company are "outside" workers whose presence in the nuclear industry is only temporarily legitimate, and only under conditions that constantly remind them that they do not belong to EDF. This is reflected in the reality of distinct entries onto the sites for the two groups, different washrooms, and vending machines and restaurants reserved for EDF agents. The companies' "outside" status is to be felt most strongly in the fact that their personnel are excluded from representative organizations (CE, CHS-CT). Even "outside" workers who are personnel representatives in their own company can only bring their prerogatives to bear in a nuclear plant via EDF personnel representation bodies for the site; there is no contractual tie between outside company workers and the site director that guarantees negotiation and representation rights.[3] This puts EDF agents in a position of near-absolute power over company workers. The fact that DATR workers are not represented in nuclear

[3]On the Chinon nuclear plant site, a site CHS-CT has now been put in place that allows for representation of service company management and employees. But it is only a consultative body, not a representative one.

site bodies has allowed for a non-negotiated division of labor and radiation risk in social terms, and to labor management in terms of radiation dose.

STRATEGIES FOR THE SOCIAL DIVISION OF LABOR AND RADIATION DOSES

Company Strategies

Accounts from employees interviewed suggest there are four types of company strategies for organizing labor and determining job assignments, qualifications, and radiation dose.

1st Strategy: Some Companies Become Specialized in Technical Operations on IPS (Important for Safety) Components

Pipe and valve operations require skilled work and acquiring know-how primarily through experience. To ensure reliability of operations requires major investment in training that is specific to the nuclear industry; above all it requires a prolonged on-the-job learning process.

This in turn means the companies need to hold on their employees, which means giving them permanent contracts with attractive enough wage guarantees, and ensuring that they remain in practice so as to get the anticipated return on investment in their training and skilling. These workers therefore must not be exposed to near-maximum radiation doses, as that could mean "putting them out to pasture" during shutdowns, a period when, on the contrary, they should be fully active.

These workers are highly valuable for their know-how, their technical pluractivity, their ability to operate on any type of valve, circuit or pipe. Companies that have adopted this strategy often broaden their operation capacity to include non-controlled zones on nuclear sites along with other types of activity. According to persons interviewed, these companies do use a few employees on fixed-term contracts, employees whose skills they are familiar with and whom they call in regularly. However, they never use temporary workers.

Pipe and valve companies that have chosen this strategy have been hard hit by reduced unit shutdown time since it is extremely difficult to reconcile responding immediately to EDF demands, particularly given the enormous need for operations during the shutdown "season," and assigning permanent employees trained in "nuclear acts" to other tasks in other sectors of the activity that involve other kinds of know-how during the months when EDF work volume is low. Moreover, shorter unit shutdowns occasionally go with a sharp fall in volume of activities for EDF or constructors, despite established partnership contracts. EDF executive managers can decide not to honor contracts, even if the companies have met initial EDF demands (see Part II). This leads to increased sub-subcontracting or labor leasing among companies, especially when EDF makes emergency demands.

Sites are always decontaminated before these companies' workers start plant operations. If the workers still deem radiation dose and contamination risks too high, they can also request that EDF bring in decontamination workers from other companies. This allows them to devote themselves exclusively to technical operations, without having to manage any "side operations," namely radiation protection. However, it presupposes another type of specialization: protecting skilled pipe and valve mechanics.

2nd Strategy: Companies Opt to Specialize in "High-Dose" Operations that Require No Particular Technical Qualifications and Have No Direct Effect on Safety

Here operations are simpler than with the first strategy, but as shown above, they presuppose high speed execution to limit radiation exposure. These are tasks that imply getting a heavy dose in a short time. Employees doing this work are inevitably "put out to pasture" part of the year, including during unit shutdowns.

Even though the state takes on the partial unemployment costs involved when stable, permanent employees of this kind cannot work, there remains a cost for the company due to these workers' non-productivity (wages, managing "unoccupied" personnel, paying employer's part of welfare protection). These companies therefore make use of temporary workers to handle radiation protection constraints. This way they don't have to pay the costs that go along with immobilized workers who have already gotten the maximum radiation dose.

These companies use temporary, temp agency, or job-by-job contract personnel, supervised by a few more qualified and experienced permanent employees, as may be seen in the following account by a DATR pipe and valve mechanic specialized in the "dose-costly" operation of opening steam generators:

> I once worked on a job where there were four permanent persons for 32 persons, one of whom didn't enter the zone because he was the foreman. That makes 28 temporary or single-job contract persons. (Jean-Louis, permanent-contract mechanic, 37)

3rd Strategy: Companies Develop an Internal Division of Labor that Ensures the Desired Flexibility for Performing the Seasonal Work Imposed by the Nuclear Industry

Companies using this strategy have both stable, skilled personnel as well as workers under individual job contracts and temporary workers who have done EDF training (DATR and QSP) and are therefore specialized as nuclear temps. These workers are only offered jobs in the nuclear industry, which is a way for the temporary agency to capitalize on their DATR training. If they are off work part of the year because they've reached the maximum dose, they end up on unemployment without any hope of being given assignments outside the nuclear industry.

4th Strategy: Sub-Subcontracting and Labor-Force Loaning among Companies are Also Ways of Perpetuating the Division of Labor and Dose, Namely for Pipe and Valve Mechanics

One company "leases" its workers to another for operations in the controlled zone during which, in what are clearly sub-subcontracting arrangements, temporary workers—called "Kamikazes" by their DATR colleagues—are assigned exclusively to dangerous operations. They appear and then disappear after these operations, which expose them to a high level of radiation.

It thus seems that temporary workers assigned to these high-exposure tasks are the most likely to receive the maximum dose and consequently be prohibited from working in controlled zones.

Subdivided Tasks and Sharply Inegalitarian Distribution of Exposure

Non-destructive inspection activity offers a good example of how tasks are fragmented into extremely small units, and the corollary of extremely unequal exposure to ionizing radiation.

Companies specialized in non-destructive inspection operate on the request of the GDL[4] a centralized EDF structure that has hired them on contract for nuclear plant operations, or as companies subcontracted by subcontracted companies which may be either constructors or mechanics companies. The employees interviewed regularly refer to how minutely tasks are broken down:

> Work in the plant is highly compartmentalized. One person takes the shot, another develops it, a third takes it back and forth, another deals with the papers, the last one checks the film. (Marc, fixed-term contract inspector with an outside company, 33)

Tasks are broken down and distributed to two types of companies: those that prepare operations, ensure they are monitored correctly, and analyze the results; and those that actually inspect the given pipe or seam on site.

This way of dividing labor is identical to what can be observed in pipe and valve work and corresponds to a particular dose distribution. The first sort of companies have permanent skilled technicians; they subcontract the on-site inspection work.

[4]The GDL (group of laboratories) is an internal centralized EDF service in charge of technical worker safety control of plant installations. The GDL is located in the Paris suburbs.

In the first type of company, radiation dose is managed by assigning workers to work posts or sites that vary in terms of exposure:

> There are jobs that don't require taking any dose. When you stay in the lab, for example. That allows you to recover a bit. It's not in the interests of our company to have us always at the maximum limit. (Antonio, permanent-contract inspector, 50)

Some inspection companies do many jobs outside the nuclear industry, and thereby manage not to have workers at the dose limit. For others specialized in nuclear, EDF has taken the initiative of using some non-shutdown time to train outside company workers:

> EDF decided to do a training program in winter once. It lasted two, three months. It was kind of an outward show on their part. They need inspection companies for unit shutdowns, but they can't offer year-around work, though that's in the process of changing. So the training is a way of remunerating the companies, since EDF pays for it, but EDF gets something out of it too because the training means you're better informed about developments in the oversight work done by the GDL. For oversight to be effective, you always have to collect your information using the same method. In fact, it's first and foremost a matter of interpretation, like in radiology. So it's important that people work the same way. (Marc)

The rest of the time, employees are on forced inactivity, but with the requirement of being immediately available. Inspectors say they are on "stand-by."

In companies whose workers do on-site inspection, the proportion of employees on fixed-term contracts is high. On the other hand, it seems that non-destructive inspection activity is always thought of as important for general safety, to the point where EDF has prohibited using temporary workers in this activity:

> For 4 or 5 year now, EDF has forbidden inspection companies to employ temp workers. I think that's for quality rather than safety, because they employ temp workers in other jobs. (Antonio, permanent-contract inspector, 50)

Managing Employment/Job Assignments, Division of Labor and Dose

In the typologies given above we see diversified types of job assigning: permanent skilled workers tend to be used for operations that are either important for general safety or highly specialized, while temporary workers with little or no job skills do site preparation jobs or jobs involving high exposure to ionizing radiation and receive no job recognition for it.

Tasks that involve high radiation doses are of course not performed by permanent, skilled workers:

> Putting in SG plugs—we did that twice and then stopped because of the dose. I did it. You've got hot spots: 28 to 30 rems, and 19 rems in the surrounding air. You go in for 2 minutes, you get 450 millirems. You've got people who do that all year round. (Guy, permanent-contract mechanic, 35)

If permanent skilled workers deem that contamination or radiation levels are too high they can demand that protection measures be put in place immediately:

> I always say it's the person himself who manages dosimetry. It's how the person works. You know the risk you're taking when you go into the zone. I've got my way of working. The only risk there is is contamination and radiation. Those are two factors you can't control. And you get them, you get them! To begin with, you do a map [of dose discharge], a site assessment, and you see what dose there is. Afterwards, we get the system changed. The system is individual protection, and you've got to cover the pipes with lead, the idea being to lower the dose discharge. The dose discharge when you take off the servomotor is around 35 to 50 millirems per hour. When the valve is open, its goes up to 60. Since these systems have to follow the ALARA system [ALARA: "As Low As Reasonably Achievable"], they take precautions. We have it decontaminated after the opening. . . . Actually EDF brings in a company that does that. If you didn't decontaminate we'd get 1.3 rems. If there's lead to put on, they're the ones who do it. We're not authorized to do that. If you need to put in a lead mattress , they're the ones who do it. (Guy, permanent-contract mechanic, 35)

Nuclear servitude workers who perform low-skill tasks to prepare for pipe and valve operations or non-destructive inspection are subjected to the combined difficulty of limiting dose in highly exposed operations, meeting time and quality requirements, and ensuring the contradictory matters of their own health (i.e., not getting exposed to above-regulation radiation levels) and their employment (not getting put "on the peg" [see Glossary]).

The ALARA principle ("as low as reasonably achievable") is part of the radiation protection doctrine taught to DATR workers in the compulsory training programs (see Part II for a further presentation). ALARA leaves wide margins for interpretation. An "achievable" level of exposure to ionizing radiation seems implicitly a function of the exposed person's place in the division of labor; it is fair to speak of socially differentiated degrees of exposure. Within the framework of regulation limits, the "achievable" level for a temp worker is much higher than for an EDF agent or a technician operating on material that is important for general safety, and this poses no problem for the owner EDF, which does not have to deal with the consequences.

Accounts by DATR workers reveal the operation of an implicit system of managing employment by dose. In accordance with this system, EDF anonymously "rotates" temporary workers on highly exposed positions.

MANAGING EMPLOYMENT BY DOSE: WHAT MANEUVERING ROOM DO OUTSIDE WORKERS HAVE?

The division of labor and radiation dose brought to light above (see box below) is reflected in the fact that outside workers are made to bear the greater part of the collective annual dose, a proportion that in France rose from 60% in 1978 to 82% in 1990. Since then it has remained between 80 and 82%. After a major increase in annual collective dose up until 1992, the level began falling in 1993. The observed fluctuations are due to several types of factors: technical (related to nuclear plant age and nature of maintenance operations), organizational (related to conditions and radiation protection measures implemented in particular plants), and economic (related to reduced number and duration of maintenance operations for cost reasons in recent years). In 1992, the collective average dose (EDF agents and "outside" workers combined) was 2.36 mSv (man-sievert) by reactor. From 1977 to 1996, the cumulative collective dose reached 1000 mSv.

THE DOSE

In this work, the common term "dose" was chosen over terms used in scientific or regulation vocabulary related to scientific definitions of the biological effects of ionizing radiation. That vocabulary distinguishes between "absorbed dose" and "dose equivalent." The first corresponds to the radiation dose actually received; the second to dose absorbed by biological tissue multiplied by an ionizing radiation quality factor. The collective dose equivalent (erroneously but commonly called collective dose) is a complex statistical relation between "detriment" (that is, the probability that harmful effects will appear) and dose equivalent distribution over a given population (see studies by the International Commission on Radiological Protection ICRP 26, 1977, and ICPR 60, 1990). Collective dose is expressed in man-sievert.

In our study of health regulation oversight (Thébaud-Mony et al., 1992), we counted average monthly doses by looking at month of "outside" workers' presence in the controlled zone (using month-by-month exposure records) and observed the differentiation in dose received by activity exercised and type of work contract. In 1988, the year the study was conducted, the average monthly doses by month of presence in controlled zone were 11 to 15 times higher for outside company workers than for EDF agents. Workers in certain jobs (heat insulators/laggers, pipe and valve mechanics, nuclear servitude workers) displayed significantly higher average doses than those in other activities. Workers on temporary contract had also received higher doses.

Other studies indicate that this trend is persisting over time. The data in EDF's DOSINAT file (CSSIN, 1995) and data collected for a study conducted directly with the companies by IPSN (Pages & Hubert, 1994) show that average individual doses are three to four times higher among "outside" workers than EDF agents. Moreover, among outside workers certain types of activities are much more exposed than others. Workers in mechanics, nuclear servitude, and inspection activities receive the highest proportion of the collective dose. Meanwhile, other study results—CEPN 1986/1988; IPSN 1994, and the DOSINAT data 1995—show stability of average individual dosimetry over time.

The dosimetric data in the STED study (Doniol-Shaw, Derriennick, & Poitiers, 1998) confirm the uneven distribution of radiation doses received by type of activity and work contract: single-job contract workers are the biggest category of workers exposed to more than 20 mSv annually. It should also be pointed out on the basis of this that geographical mobility from plant to plant is associated with high dosimetry for a large proportion of relevant cases.

Certain results point particularly clearly to division of exposure and type of employment contract by dose. In highly exposed activities, there is a category of worker extremely exposed to high doses that reflects the division of tasks described above in which certain employees are called upon to undergo the greatest exposure so that technicians can perform important-for-safety tasks. There are also temporary workers who undergo high exposure for very brief moments in greater proportions than all fixed-term contract workers. Three workers, two of whom were on single-job contracts, received a monthly dose of over 30 mSv (i.e., well over the current quarter annual limit of 20 mSv). They say they worked less than 15 days on the site. Lastly, workers classified "other" in the STED study also had higher monthly dose levels that workers in classic types of activity in the nuclear industry (the jobs done were not specified). Moreover, it should be underlined that the highest average monthly dose levels are correlated with a lower skill level, uncomfortable working conditions, an accumulation of exposure to risks of various types (toxic chemicals, dust, heat) and heavy time and work-pace constraints.

On the other hand, there are no studies providing information on dose received by work post, or on changes in work posts and exposure; i.e., there is no longitudinal approach to individual dosimetry by "outside" DATR worker's occupational trajectory. The fact is that only such data would be able to provide quantitative and qualitative information on what exposure workers really undergo, and how exposure is managed in terms of alternating employment and unemployment, not only in terms of average doses but by characteristics of posts occupied, organization of labor, and type of employment contract.

The data cited above nonetheless allow for apprehending the structural dimensions of the social division of exposure to ionizing radiation in the worker population of the nuclear industry (EDF and "outside"), and they substantiate the hypothesis that type of employment contract is determined as a function of exposure distribution. Part II takes up the question of the management instruments used by EDF and subcontracting companies. First, however, come accounts by "outside" DATR workers themselves that will allow us to understand how employment and exposure are managed. In the day-to-day organization of maintenance work, the division of labor and dose determines DATR workers' maneuvering room in each of the types of jobs studied, not only with regard to exposure to radiation but also to keeping their jobs.

Pipe and Valve Mechanics

From the skilled worker who machines a part after determining radiation discharge and requesting, if necessary, that the worksite be more thoroughly decontaminated or that a lead mattress be furnished (lead is the only effective barrier against radiation), to the temp worker who refers to himself a "rem beast" or "rem meat," all situations may be found in pipe and valve mechanics, from ample maneuvering room for managing individual and collective dosimetry to near-absence of maneuvering room, characteristic of the worker who knows he is there to perform "dose-costly" operations (opening tanks, inserting SG plugs) to prepare for other maintenance operations.

Describing the operations enables us to understand how exposure to ionizing radiation and the consequences of that exposure on outside company employees' jobs are managed through the social organization of maintenance work.

> Let's take the example of a plant where there was a problem two years ago. We were working on a primary cooling circuit valve [generally highly radioactive] that was discharging a dose of 43 rems. For that we had to see the internist immediately. Given the dose level, I really don't know how many of us were going at that valve but it must've been really a lot of persons.
>
> And what happened?
>
> Didier: We were doing check-up. But I've never seen a valve with a dose discharge like that. That plant, really you could almost say that you were in the orange zone as soon as you entered the reactor building. It's death. That unit, I mean. It's the one that's had the most cracks in the cladding.[5] Let's not say we received an alarming dose because everyone went through the same thing. I took the screws off the valve, also the lid and the seal retainer—that's all. I spent a quarter of an hour at it—took maybe 400 millirems [1/5 of the maximum annual dose]. I didn't return to the zone after that, I went into the secondary circuit. After that, the persons in secondary went to work on it because they hadn't planned on such a dose. That's good. They might've said "You take a dose up to. . . ." It was really well planned, so the dose was distributed well. With a dose discharge of 43 rems it really flies, and you just can't get stressed out. Guys who get stressed out look at their dosimeter instead of their work. That's why you can't put just anyone on the job. If ever I saw that the person was getting stressed out, I'd have told him to leave. It's already makes no sense to be two people

[5] When nuclear plants age, one sign of wear is cracks in the fuel cladding. This is considered a major source of radioactivity in the circuits.

> on a job that only requires one. In any case, you've got to have your head on your shoulders, you've got to say "the faster I go the less dose I'll get."
>
> Guy: Not really. It's not "The faster you go the less dose you get"; it's "If the job's been well planned out, it goes really smooth."
>
> Didier: Yeah, that's it. I knew exactly what I had to do before entering the airlock, I knew exactly where the key was, where this or that was and everything.
>
> Guy: You knew where you were.
>
> Didier: I knew exactly what had to be done and the time I was going to be working on it. You've got to have the operation clearly in mind. You can't go in as if you saw the valve already taken apart. For us it hasn't been taken apart yet. We take it apart gradually.
>
> Guy: You've got to have prepared everything. If you get as far as the valve and you're in a Muru suit and you say "Darn, I forgot this or that," you've got to get undressed again and all that. . . . You have to be familiar with the material you're going to be working on. Plan everything in detail beforehand.
>
> Didier: I think that when you work in the controlled zone, if you've never worked on a valve or a particular type of material, you should say so. And if you've never done it, you go see the EDF contact person. And it's preferable to refuse to do the valve than say, "I'll do it." The guy who does that is going to have a hard time, he's gonna get seriously dosed, and that'll give the company a bad image. If you don't know how to do it, you say so. Instead of screwing up. If you don't know, you don't do. (Didier, permanent-contract mechanic, 26; Guy, permanent-contract mechanic, 35)

This dialogue shows the importance of being familiar with the operation and being able to anticipate how it will proceed in order to control exposure to ionizing radiation. Experience is the only way.

Still, the beginning of the learning process involves subdivided tasks as well as radiation dose "sharing" between experienced and temporary workers.

> In the beginning I worked on the four aces—that's four valves that form a square.[6] That's where it spits the most in valve work. If you're well organized it's ok. If you're not, it's not. Because you're not just two or three persons doing it, you're going to be several teams. The dose discharge is so high that you've got to change people really fast. First comes a cleaning team—you get out of the way. Then a dismantling team, then a team that comes in to take out all the internal parts. We move away—you're not going to hang out near there. You go clean up at the "hot shop" where it isn't spitting, then afterwards there's a reconstruction team. It's organized, I'll give them

[6]Reference is to four big valves in the reactor building, usually highly contaminated. Important resources must be used to ensure radiation protection for maintenance workers operating on these valves.

> that. It's about one of the only things they follow carefully. (Eric, permanent-contract mechanic, 22)

But one needs to have a certain ability to maneuver in order to get assigned outside the controlled zone when dose level is high. Temporary workers whose contracts do not go beyond the particular job do not have this ability, and for them getting a high dose means the end of their contract. A young valve mechanic who had received 1.5 rems in a few days' time was terminated.

This is the kind of situation in which "cheating" practices develop. Cheating is when an employee doesn't wear the dosimetry apparatus for fear of losing his job. He doesn't want high doses to show up on his record. Employees are particularly aware of the danger to their jobs now that EDF has set up DOSINAT monitoring,[7] which has led to excluding workers from the site if they have exhausted their individual dose "credits":

> We're allowed 5 rems a year [by the law in effect at the time of the study]. We get dosed. You can't help it because there's a lack of personnel, only two guys in places where there should be four or five. Now there's a computer program they're starting to use at the plants.[8] If you get such and such a dose here, everything is added on the computer. If you go to another plant and you're at your dose limit, they won't let you in. EDF set this up.
>
> The big companies with stable personnel and a lot of different operations can alternate between zone and non-zone jobs. But for "leased people," temps, there's no more work for you if you've gone over your dose, and this explains the cheating when it comes to wearing the saphymo [name for the electronic dosimeter], so you won't be prohibited to work in the zone because of your dose. You've probably seen guys who are in the leasing companies, after a unit shutdown, when they've reached the [maximum] dose and they don't know what to do, when they're in danger of getting stuck in the house for two or three months. They're going to be thrown in a drawer and when there's no need for them anymore they'll pull them out of the drawer. And that's real. Why do some guys take off their saphymo to work? So no one will know the dose they've gotten, so they can work longer. I know guys who've done that. (Alain, permanent-contract valve mechanic, 26)

This also affects workers threatened with partial unemployment due to high exposure:

> It's during very short operations that you get dosed like that, it's specific to the steam generators. We spend half an hour getting suited up, we work for

[7] The DOSINAT system records individual dosimetry "live" on the computer as it changes by means of an electronic dosimeter. Dose is recorded for the individual worker by name, and the system covers all sites. This allows EDF to know at any moment the total dose of any operator.

[8] Speaker is referring to the DOSINAT system.

> one to three minutes, and after that it's over, we spend the rest of the day playing cards. . . . If we go beyond 5 rems, we're going to get put out to pasture. May be paid partial unemployment: 27.50 francs an hour (€4.20). Sometimes you can go into the zone to do some little thing, but . . . They call us "rem beasts," if you really want to know. And then there are those who plug up the holes in leaking steam generator pipes—they're called jumpers, and they get dosed even worse than we do! The SGs are tubes, heat transmitters. The tubes get worn out and some leak.
>
> Who handles that?
>
> Other companies with lots of temp workers. (Jean-Louis, permanent-contract mechanic, 37)

According to the comments collected, and in the absence of legislation protecting the employment of workers who have reached the 20 mSv annual limit, EDF's lowering of maximum dose levels for plant entry can only push companies to increase the number of temporary workers and encourage employees not to wear the dosimeter during "dose-costly" operations:[9] "That's the problem for leased guys," as the mechanic cited above would put it. In the judgment of another, lowering the maximum dose to 2 rems (20mSv) doesn't make sense given the increased radioactivity due to nuclear plant aging:

> It's impossible given the condition of the units. Yesterday I saw a guy in servitude get it again. After reloading the fuel, they go through with a machine called the Véda, a little submarine that goes under water with brushes and cleans the walls before the tank is emptied. There was a contamination problem, so the guy went in. He got 480 millirems in 20 minutes [nearly 1/4 of maximum annual dose according to the new regulations]. If he gets that three times a year he won't be able to work anymore. The plants in France are too contaminated. (Patrice, permanent-contract mechanic, 27)

> And it's not saying 'exposed' instead of 'contaminated' that's going to change the problem.[10] That's right—EDF doesn't want us to say we were contaminated anymore. (Guy, permanent-contract mechanic, 35)

[9]On February 3, 1997, officials at the Saint-Laurent-des-Eaux nuclear plant announced to the press that two employees of the Jeumont Industrie company were no longer eligible to go into the reactor building B1 controlled zone "because they were not wearing one of the two apparatuses required for dosimetry" (*Le Monde,* Feb. 5, 1997). The article does not give the employees' version of the facts and does not indicate which of the two recording apparatuses was not being used. It is nonetheless highly probable that it was the electronic dosimeter that allows for checking the dose level of any DATR worker at any moment.

[10]Nuclear installation executive managers have tried to substitute "exposed" for the usual term "contaminated" in order to attenuate the analogy in general representations between radioactive contamination and infectious diseases such as AIDS or tuberculosis.

These few examples show that when it is impossible to avoid human intervention in highly irradiated nuclear plant areas, exposure of valve mechanics is managed by assigning them to jobs as a function of dose level. This is based on two complementary strategies:

- Determining job assignments in such a way that each individual worker stops short of reaching the regulation limit. Companies that have invested in the training of skilled permanent workers try to assign them to a combination of controlled-zone jobs and classic maintenance work inside and outside the nuclear industry. Low-skilled workers on the other hand become what are called "rem beasts," and given the dose limits and practices of exclusion from the site for those who have reached those limits, they are only hired for short-term contracts or on a task-by-task basis.
- Day-to-day managing of individual dose levels by employees themselves and the company. In companies with a high proportion of permanent-contract workers, dose levels are managed in this way, which allows for planning operations at the level of teams and the overall level of the set of operations to be done. This type of management is done without taking into account the DOSINAT file that connects the different nuclear plants. The employee himself notes each day's dose and transmits to the company all information on collective and individual radiation received during a given operation. This information in turn determines which workers are assigned to which sites. In this way, exposure levels are known to both employees themselves and the company project organizer.

This composite strategy can only be efficient, however, if workers have a permanent-contract and stable income, and therefore feel it to be in their interest, in the interest of their health, to have their radiation exposure levels known and taken into account throughout the year. Workers "specialized" in high-exposure operations and in danger of losing what is already not very steady employment, especially those who have only seasonal or job-by-job contracts, are likely not to want their high exposure levels known.

Non-Destructive Inspection Workers

The examples collected from non-destructive inspection workers show variable types of exposure to ionizing radiation and greater or lesser maneuvering room for workers in this area.

- Dose received varies over time and by job assignment strategy. When there are numerous operators (which in fact implies an increase in temporary workers), the dose is "shared":

> Inspectors sometimes get high doses. At the moment I'm at 1 rem for twelve months. Two years ago I was at 4 rems [twice the maximum according

> to the new regulation]. It's going down a bit, in part because there are more of us. That allows for sharing the dose. But it depends on the sector you're working in. (Antonio, permanent-contract inspection worker with an outside company, 50)

- Dose varies by work post and EDF arrangements for reducing exposure:

> It isn't using [radiation] sources that exposes us the most[11], it's a function of the zone you're working in. If the guy does his job well, if he follows all the rules. . . . I mainly inspect steam generator seams. If it's done in water it's okay, if it's not, we get dosed. Regulations say we're only supposed to work with water. But it happens that we don't work with water. Personally, I've always refused. My health comes first. (Jacky, inspection worker with an outside company, 50)

Another inspection worker remarks on EDF's attempts to robotize operations:

> I think they're studying a machine—a robot that's actually already working. They haven't finished the fine-tuning, but the results show that you can in fact use robots to inspect the inside of the capacities. They're moving forward on it gradually, but really, things have got to change. It's not easy, but they've got to understand—we can't keep on like this. We're willing to make an effort, but they have to do likewise on their side. (Antonio, permanent-contract inspection worker, 50)

- Dose varies by how old and "worn" the plant is and how contamination is managed at each plant. Antonio speaks of the difference between new and older plants:

> In the space of three weeks early last year, and for the same operation, I got 200 millis at Cattenom and 1 rem at Bugey. For now there's nothing to be done about it, they just haven't found a way to clean the old plants. But they've got to find one, because it's really going to be a problem. At Bugey they did just about the same operation we did at Cattenom. For a given job there were three of us, we must've each got 30 millis, whereas there were two of them and they each got 500 millis. The problem was radioactive mud deposits at the bottom of some piping. You got about 40 rems on contact, I think. It seems EDF is working on a robot to clean that kind of mud. (Antonio)

There is clearly a contradiction between shorter shutdowns and increased exposure intensity due to plant aging:

> In the matter of dosimetry, the shutdowns are shorter but the doses are increasing. I recently did an shutdown at Bugey, for example. I spent fifteen

[11] It will be remembered that the instruments used by inspection workers are themselves radioactive.

> days there. It's an old plant with high dosimetry. In fifteen days I got 500 millis, the equivalent of a monthly dose [a three-month dose by the new regulation]. Since the shutdowns are shorter, we work more sporadically so the dosimetry is more spread out. Now they're likely to give you an x-ray exam after a few days, whereas before it was after a week on the job. So we get both less and more. It seems to me there should be more inspectors—so that there could be lower doses. There are even circuits where they do inspection by asking questions because the dosimetry levels are so high. At Bugey, for example—but maybe that's a special case, I don't know—four or five people got contaminated internally during the shutdown and twenty or so went to the infirmary for contamination. They posted it. Going to work under those conditions, knowing there are accidents like that . . . we do it because you've got to work, but really, it's no picnic. (Marc, fixed-term contract inspector, 33)

Permanent inspectors say they "won't go" when the danger is too great or if they've already "been dosed." Having the option to refuse depends on how sure an employee feels that he won't lose his job for it.

The intense radioactivity of aging plants causes more than high dosimetry levels over a very short time; it also leads to external and internal contamination. Marc wonders what becomes of recorded dosimetry levels and why they don't seem to lead to lower exposure levels:

> For dosimetry, for example, given that there are people who get big doses in very little time—that is, during an eight-hour operation [on some circuits the dose can get as high as 400 millirems, 1/5 of the annual dose according to the new regulations], the question that comes to mind is . . . I mean for the shutdown it's noted by the project heads who do the dosimetry sheets for workers, but we don't know what happens to all that. Because in fact there hasn't been much of a change. You could almost say we're at the same rates as ten years ago, except that the operations are shorter. For dosimetry levels, nothing's being done. You wonder what the dosimetry sheets lead to. It might seem they're not really being used. (Marc)

Lastly, plant aging leads to an increase in the area of plant space contaminated by radioactive particles, and this causes contamination and irradiation even in closely monitored zones:

> After ten years in operation, a nuclear plant's going to have radioactivity in spots that shouldn't be radioactive, like in the auxiliary building. Normally those places are clean. But after ten years, it starts spreading. They find sources in the piping that shouldn't be there. It's getting serious.
>
> And there aren't systematic inspections?
>
> Usually not, because it's not supposed to be radioactive.
>
> Is this in the zone?
>
> It's in the zone, but in a place that was never supposed to become radioactive. No, in fact, in the auxiliary building there are two or three

> places that aren't at all dangerous. You go in, you come back out again and your apparatus is still at zero. But they realized that because you kept going through there, each time you were getting a little. So they started asking the question. They asked it so well that they found a source there that had no business being there. (Jacky, inspector with an outside company, 50)

Radiation protection of "outside" workers in the plant does not always seem a priority:

> What I find really disturbing in nuclear plants is that when you ask for assistance from the radiation protection team, they take their time coming to check a place where it's spitting . . . whereas the work has to be done.

Contamination is tending to become an ordinary phenomenon, while irradiation is considered the real problem in aging plants:

> I've been contaminated externally. The biggest was 3000 impacts to the head, but that got washed off in the infirmary. It also happened to my hand. That time I had to leave the plant with my contaminated hand. There had only been about thirty impacts, but with all the scrubbing there was danger of bleeding and it becoming internal. So the nurse put some vaseline on and taped on some vinyl real well. My hand sweated and everything came unstuck. Two days later they took off the glove and everything was fine.
>
> What happens when you get contaminated?
>
> You change suits, you put on special boots, you go as far as the first detector gate and you take everything off. When it's just the clothes it's not too bad. That happens often, nearly every day. You rub up against some piping—it's inevitable.
>
> And what are the effects of internal contamination?
>
> Everything depends on the radioactive element and the contamination degree. Not long ago I got slightly contaminated internally at Bugey, from iodine. It wasn't a big deal. There must have been about 20 impacts to the head. After eight days there was no problem anymore. But it was my fault in a way. When your head's been contaminated, you can't decontaminate yourself, the nurse has to do it. You can only decontaminate your hands, legs, and feet—not the head. There's always a guy assigned to the changing room. He says to me, "There's no one here anymore, and in fact you don't have much of anything." So I washed myself, and it must've slipped a bit into my nose and mouth. I went to get checked and I had a little iodine. It wasn't much, but if I hadn't washed myself that wouldn't have happened. . . . Contamination's not the real problem, it's irradiation. And that's something you can't . . . I mean you can always diminish the risk a little, but there are times when you can't do anything. And it's going to become a really worrisome problem. Some plants are not even old yet and there are already "hot spots" in them. When you get 40 rems on contact with piping in a plant that's only six years old. . . . They've got a life span of about 20, 25 years, maybe 30. If you're already getting 40 rems at six years, that's really

> a lot. Even at Cattenom, which is brand new, when we did the first shutdown it really started spitting.
>
> What's that due to?
>
> Badly filtered radioactive deposits, maybe. We don't really know. (Antonio)

- Variability may be associated with the contradiction between productivity and safety demands.

Non-destructive inspection workers are acutely aware of the contradiction in their work between the safety constraint and the profitability concern that forces them to work at night and constantly reduce operating time:

> This is going to seem ridiculous to you, but you get accustomed over time, and I realized, for example, that last year I hadn't even noted my dosimetry levels. But I know I had a slight source accident in October. Carelessness on my part. It was night work and I was tired. We had been working a lot that month. And I stayed too long next to the source. I got 330 millirems. (Marc)

To this are added constraints linked to the nature of the agreement concluded between EDF and the company. Marc explains that one of the problems for company workers is the fact that the contract is drawn up not on the basis of service performed but number of films used. These films don't record worker dosimetry in the controlled zone but are used for "x-raying" the piping. As mentioned, workers use radioactive iridium sources to do this—sources that they must also protect themselves from. The cameras are heavy and they have to be careful to stand far away from the source. But the farther away they stand, the longer it takes to prepare the source.

> There's a remote control with an ten-meter long arm, and on the other side there's an ejection tube that's about three meters long with a radiation nozzle or a collimator. If the person is doing production, they'll most likely not use the collimator, because it's also made of depleted uranium. It's very heavy so it's difficult to manipulate, you lose time. You have to keep it in place with magnets or bungee cordstime spent getting set up that hurts profit. If you respect the rules of the game, you don't get that much radiation. (Marc)

Variation in dose level and exposure frequency is implicitly taken into account in managing worksites, job assignments, and dose. For the workers in question there is no official negotiating room. Workers negotiate only as individuals, and the lower the skill level the less maneuvering room a worker has. As for working conditions, nothing is negotiable. "Dividing the dose" involves spreading it out over a large number of workers—meaning temporary workers. Maximum exposure level is the crucial cut-off criterion for letting a worker enter the plant, as nuclear servitude agents readily point out.

Nuclear Servitude Agents

"Getting as little radiation as possible" is the dominant preoccupation in individual and collective work organization for these agents, whose function, it will be remembered, is to intervene precisely where dose discharge is particularly high—where"it spits."

> Decontamination is done after loading and unloading the reactor core. We decontaminate to prepare for the others' operations. We're in Muru suits on our knees in the tank, scrubbing and mopping. At the bottom of the pool, it spits at very high levels. With the beam (a kind of robot which allows you to draw water at 600 bars) you can drain "hot spots" in such a way that guys don't have to go over there. But management won't let us use the beam; they prefer to send guys in. These are operations where you can get 800 millirems at a go. We're there in Mururoa suits with our scrub-brushes right where it spits. I can only get to sleep with sleeping pills or a tranquillizer. I got my pancreas operated on. Now I'm on Temesta [tranquilizer]. During operations like that one I can't take my tranquillizer. I only sleep about three hours. In three months I got 1.4 rems. As site foreman I should be supervising, but since we never have enough personnel, I work with the guys. But if the big exposures keep up, I won't be able to continue. I'm thinking of going to the ANPE [state employment agency]. (Patrick, permanent-contract nuclear servitude worker, 35)

Since this work directly involves "getting dosed," the only maneuvering room that remains—for companies as well as workers—is to cheat when it comes to recording dose received. DATR workers do not always willingly leave behind their dosimeters; in some cases they have been pressured to do so by their bosses, pressure that includes the threat of being fired:

> What do I do? I go where it spits. We're "rem beasts," if you want to know. Bad things go on and no one will ever know. It goes on at night when no one's there. . . . There are some conscientious companies, you have to admit. Right now I'm working for a serious company—I won't tell you which. But there are others who don't give a shit. In 1990 at C. I was working in the middle of the night. There were no SRP agents. I was asked to throw away my saphymo and go get double dosed. I refused and I got fired immediately. That's the way it is. (Jacques, limited-time contract nuclear servitude worker, 33)

In this context, lowering the official maximum radiation levels for working is experienced not as a protective measure but a threat to employment and a reason to cheat more often in dosimetry recording.

> Will the fact that the maximum dose has been lowered to 2 rems change anything?

There will be more cheating and more people unemployed. They should make subcontract companies hire people on permanent contract. But that'll never happen.

Have you ever been at 5 rems a year?

Never—I know how to take care of myself. But the young guys just out of school, they're made to take big doses, they don't know anything. Four days of training is not enough. The QSP? It's a fantasy. There's a serious lack of training.

In connection to job assignment by dose, a young permanent employee of 23 shows a concern for protecting his health as well as keeping his job:

When I work, if there's a place where the dose is really high, I don't go there. I go see the boss and ask him to put in a young guy who hasn't been dosed. Because they take these young guys in mid-year. We've been getting dosed since the beginning of the year, whereas they're brand new. Send them in, let them get dosed, let them get to the same level as us. Because afterwards they're going to put us on the peg and replace us with young guys who don't know the job. I was 600 millirems away from the maximum of 5 rems per year last time. I chose to leave fifteen days before the end of my contract: you can get 600 millirems in no time and I didn't want to take that risk. The boss was not happy, but I don't care. . . . I've never really looked into it, but I don't think that it's too good for your health over the long run. (Hermano, fixed-term contract nuclear servitude worker, 23)

In 1994, after quitting work as a permanent DATR, Hermano says:

You've got to take the initiative yourself on site to avoid getting too big a dose.

As monitor of entries into the airlock leading to the controlled zone, Hermano observes that some workers do not follow the prescriptions, thus forcing him to call in the EDF SRP agent:

I knew some guys that went down to decontaminate the tank without film, without a Geiger counter. They didn't last long. They were specialists, high up, site project heads.

Why didn't they use the film?

Because they had high dose levels. They must've been at 4 rems or more and they knew they still had a fair amount of work to do.

Radiation exposure is a source of concern for Hermano on two points: the effects of ionizing radiation on his health, and the fear of reaching the limit and losing his job.

You're always kind of scared in a nuclear plant. You don't think about it, you don't show it, but you've always got some fear. There are times you imagine things, you get a little crazy, you think it's all going to explode.

Do you talk about it among yourselves?

Always. We laugh about it! That reassures us a little. Sometimes there are incidents, cracks in the cladding. That's never happened to me, but it happens. Then they straight out close the plant. I've never been all the way to the maximum dose, except this year. This year I had 4 rems. It had to happen!

Does it worry you to be constantly exposed to radiation?

It worries me all the time. There are times when I look to see if I'm not losing my hair. I go to the dentist regularly. It worries me all the time.

CONCLUSION

The characteristics of division of labor and individual and collective management of radiation dose studied in this chapter show institutionalization of an increasingly tight strategy of dividing up radiation exposure among a high number of operators during high exposure operations. Accounts from "outside" DATR workers show that at certain work spots, exposure is constantly high. With the aging of nuclear plants, radioactivity is increasing and spreading.

Increasingly tight surveillance of maximum dose has brought about a broader distribution of radiation among workers. To manage exposure in work spots that involve heavy radiation discharge, a great number of workers must sometimes be made to "go through" in a brief span of time so as to divide up the collective dose without exceeding regulation limits for individual workers. The complement to this strategy is the absolute cut-off point, experienced as a kind of guillotine. The workers say this works because of the new computerized dosimetry system. Once the maximum regulation dose level has been reached, the worker is of course prohibited from entering any French nuclear plant. EDF delegates to companies the responsibility of handling the effects of this prohibitive measure in hiring and wage terms. Permanent workers are put on partial unemployment; the measure thus hits their income level. Temporary workers, however, may suffer two kinds of damages: they lose their job and in some cases all income.

Ultimately, then, it is up to the DATR worker to choose between his job or his health, to choose either "freely," by leaving his dosimeter at the door, or under pressure from an employer attempting to circumvent radiation protection rules and their effects.

This is therefore not a residual situation that could be improved by better training, but a structural division of labor among various employees, who are either protected because their operations are important for safety, or there to get dosed precisely so that the first group doesn't. EDF has not had to negotiate the way this division of labor works; on the contrary it has been able to impose it by means of anonymous computer technology.

The bureaucratic character of the system and the fact that working conditions are non-negotiable can produce such heavy constraints on the individual that he himself contemplates how to slip the grip of the dosimetry surveillance system. This of course can only work to the benefit of the work organizers: there is no longer any trace of dose level. It would seem impossible to apprehend quantitatively how widespread the phenomenon is given the opacity created by subcontracting relations and labor flexibility.

CHAPTER 3

Triple Flexibility

In the organization of labor, "managing urgent problems" is today the leitmotiv. Each minute of production time costs money. Each task is an emergency. Competition demands that machines be at the top of their performance and defines production pace and the indispensable functions related to it (upkeep and maintenance). Maintenance workers live under constant pressure because they have been forewarned of what immobilizing production in a nuclear plant during a unit outage would "cost" the principal, EDF—as if, in accordance with the ideal that inspired the conception of the nuclear plants in the first place, those plants were supposed to be able to function non-stop, without maintenance, upkeep or cleaning.

Such a representation—which dissociates production and maintenance, and in fact sets up an opposition between them—has been internalized by the "outside" workers of the nuclear industry. They are in a constant state of emergency, and are haunted by the fear of not meeting deadlines. In terms of "production loss" for EDF, *"nuclear plant outages are very costly: a 24-hour interruption comes to about 6 million francs [approximately 915,000€], or even more"*. Given the uncertainties (which have never been resolved) concerning the eventuality of extending the contract between their company and EDF, they know that any delay in carrying out the work could jeopardize obtaining a new contract (for non-respect of the deadline).

This *"production urgency"* (Thébaud-Mony, 1994) validates the triple flexibility on which the organization of labor rests:

- flexibility of jobs, made possible by the social division of labor and exposure to ionizing radiation together with the use of sub-subcontracting, as described in the preceding chapters;
- flexibility of daily, weekly and annual work schedules;
- flexibility of family life, which, counting on the traditional gendered division of labor within the family—"men without families, women without work" (Cf. *Le sexe du travail*, 1984)—puts these male employees in a situation where they may be called upon at any time and sent anywhere.

As a result of this triple flexibility, any form of collective or labor union organization is difficult.

JOB FLEXIBILITY

The previous chapters showed how the division of labor and qualifications is a function of the imperatives of safety, maximum authorized radiation dose levels and the constraints of the power plants' ongoing productive capacity.

Job flexibility within the framework of this division of labor has two complementary facets.

Temporary Seasonal Employment

In relatively unskilled work, which for some signifies high dosimetric levels, the seasonal character of operations and the obligation to distribute radiation dose among employees lead the companies to resort to single-job contracts and temporary work. Daniel (valve mechanic) explains what this meant for him:

> For the first two years, I worked on a temporary basis. I found the job through the ANPE [state employment agency]. They offered to pay for a training period, and I took it. My first worksite was Chinon. I had a contract (that lasted between one week and two months). I signed a contract when entering the premises. These were short, renewable contracts, but the end of the contract was never stipulated. They fiddle with the law. These contracts last only as long as the work on the site.
>
> As a temp, you only find out on a Friday or Saturday that the next Monday you have to be in such and such a power plant, sometimes very far away. You're obliged to be mobile if you want a little more money or a somewhat more decent salary. I always worked for the same company as a temp. It was a small business that leased personnel from a bigger company. When I was in training, they told me that the temporary jobs were well organized. I had 15-day contracts. Then I got the ASSEDIC [unemployment insurance payments]. They say "We'll call you." You wait 15 days, a month, a month and a half, 4 months, the whole winter. I was registered with 6 or 7 temporary employment agencies but they're really "meat sellers". The only agency that offered me work was the one that did nuclear power plants. As a temp, I passed my DATR in the company that employed me, not at the agency.
>
> It's very simple to set up a temporary employment agency. All you need is a bungalow and a secretary. If you know somebody at EDF, it's very simple to get your hands on the jobs. That's how I set up my company. Some companies do a lot of subcontracting. The small companies have to wait, they're at your beck and call. They're ready to do anything for a contract. There's a lot of corruption. As to the accreditations, all you need is a signature.

> The guys at EDF didn't know I was a temp. You don't tell them, you say you work for such and such a company. Also, there's a bad feeling between EDF and the companies. There are sanctions. If they don't like a person, he's out fast.

Temporary workers and workers under contract thus quickly become "loyal" employees and the companies summon them according to their needs. Hermano (nuclear servitude worker), who works on single-job contracts, explains that it is necessary to be completely available when you are unemployed, so as not to take the risk of missing out on a job:

> They phone you [the company] and say: you've got to be there tomorrow or day after tomorrow. That's pretty short notice, isn't it? They only tell you at the last minute—for example on Monday for Wednesday or Thursday.

Careers do not evolve much and salaries remain low. The employees talk about the lack of career advancement due to this contract system, which is typical of managing job assignments by dose:

> There's no career plan! You only work through contracts. They give you a three-month contract, you absorb three rems and they throw you out. (Jacques, nuclear servitude worker)

Between two contracts, Jacques is on unemployment and sometimes does odd jobs. He hasn't moved around from plant to plant since he got married. His salary shows it. He earns 5,400F [€823] a month, with a "ventilated air" bonus of 103F [€15.70] per day when he works with a respiratory assistance suit. He would prefer to stop:

> Because nobody here is able to tell me what all this is going to do to my health. Yesterday you were allowed five rems, today it's two. Does that mean it was dangerous before?

As mentioned, reaching too high a dose level puts an end to the contract, which was open-ended to begin with.

"Long-Term Reduced-Time Unemployment Compensation" Arrangements

Permanent and skilled employees have no choice but to accept combined arrangements of annualized work loads alternating with periods of forced idleness. The *five-year law of December 20, 1993, relative to labor, employment and occupational qualification* permits companies to use "*long-term reduced-time unemployment compensation*" arrangements, which amount to 1 200 hours over an 18-month period—39% of legal working hours. Payment for those periods is guaranteed by the state, the company and UNEDIC [Union Nationale pour l'Emploi dans l'Industrie et le Commerce]. During the time salaried employees remain idle, they receive 50% of their basic hourly wage per hour lost, a legal indemnity that is supposed to be supplemented by a specific system of benefits that partly makes up for the difference (Le Cohu, 1994). "*Long-term reduced-time unemployment compensation*" is an extended mode of implementing the partial unemployment arrangements that were being applied by nuclear industry maintenance firms before the December 1993 law.

In fact, to be able to handle the labor-intensive short-term periods (the seasonal phenomenon) without having to pay overtime, which is more costly than normal working hours, companies whose employees are on permanent contracts work them hard during the "unit shutdown periods" (i.e., from April to October), then give them a paid holiday for a period equal to the amount of overtime put in. They thus "save" on amounts of partial unemployment during the weeks of recuperation, to which must be added yearly vacations and off-days not taken when employees are moving about.

> We'll soon become seasonal workers. There're going to be 14 unit shutdowns in August and in winter there's no more work anyway. . . . They pay us 5 hours overtime a week. We're on a 39-hour schedule. We work 44 hours, we're paid 39. They stash away our 5 hours. If there's work in winter, the overtime is paid more. If there's no work, they use up all the hours we stocked up during the year—sometimes we've stocked 200 hours, that's more than a month. Afterwards, when there are no hours left, we get vacations, and when there are no more vacations, the boss pays us from his own pocket. But right now, he's looking pretty sour. In winter he's got money because EDF pays three or four months late. We work during the summer, until September or October. He gets paid and then he pays all the guys until January. In February he's broke, since he didn't work in November and December. So then we start over again in February for the whole year.
>
> They're suggesting we go abroad, if we don't, we're going to become seasonal workers. That means you're only allowed to work during unit shutdowns and you get three years lay-off compensation. (Eric, permanent-contract valve mechanic, age 22)

The periods of forced idleness do not allow employees to dispose of their time as they would like, because, to use the inspectors' expression, they are often on "stand-by"; they can be called back at any moment to respond to an emergency order from EDF and must be ready to get up and go, even to a plant on the other side of France.

While employed, they are haunted by the idea of having to increase their unemployed time if their dosimeter level goes over the top. "*Being put on the shelf or out to pasture*" inevitably means losing income. Clearly, being put on partial unemployment or short-time work, as illustrated in the preceding chapters, is another way of managing employment by dose.

For the employees, this sort of labor organization amounts to not being able to choose one's vacation time and therefore not being able to fulfill one's family obligations:

> You tell them the dates you want for your vacation 6 months ahead of time, so as to be able to reserve a house. They don't say yes, they don't say no. And then they tell you, "You've got to be a civil servant to reserve 6 months ahead.
>
> When I was on the road, I said to the foreman on the site, "My 14-year-old has got a few problems. Can I stop moving around a little, just a little?" From then on they thought I was lazy! . . . My wife has to be both mother and father to the kids. (Patrick, nuclear servitude agent)

And the financial loss is considerable:

> We work like crazy during the summer. And in winter we're out of work. But we can't go on vacation because we don't have any money. The guys who're on the road are all getting money taken out of their salary by the bank. As temp workers, they count too much on their jobs away, their job periods. They've got mortgages to pay. They've got to feed their wives and kids. Many of them are having money withheld. They don't always talk about it, they're not proud of it. But they count too much on their job periods. In my case, I'm staying with my mother. But some of them have mortgages, they pay 3000F (€457) for rent. Come winter, you've only got 3000F left to live on out of a salary of 6000F. You've got the electric bill to pay, the car, etc. . . . (Eric, valve mechanic)

Making unit shutdowns seasonal has also led a number of companies to lay off part of their permanent personnel and "lease" employees from other companies—and to exceed the legal amount of partial unemployment time:

> There were some lay-offs at T's this year. They laid off the loudmouths, without bothering to look into their family situation. One of my colleagues had just gotten married. Last week he committed suicide. He was 32. His wife was in the hospital, expecting. And for the first time there's been partial unemployment too (1700 hours spread over the year). The crisis is because of EDF. We used to work from February to December. Now, it's March to September. That's leaves five months of nothing. So they send us off to various places, they hire us out to other companies. It's not clear how to manage that. But EDF likes that sort of policy. What does it do to work quality? I don't know. I suppose they're waiting for there to be a problem before they do anything. (Bernard, permanent-contract electrician with an outside company, 35)

This way of managing worked hours is in fact a strategy for reducing the length of working time and thus lowering payroll costs, a way to avoid paying overtime and force employees to accept annualization. This kind of strategy transfers the financial effects of the incompressible radiation amounts received during maintenance operations to the employee. Given the status and constraints of labor union negotiations for regulating work hours and conditions, as well as salaries and legal holidays, EDF couldn't ever impose such practices on its own employees. But by subcontracting maintenance, EDF has been able to impose a correspondence between paid work hours and amount of radiation exposure on outside company employees, without having to negotiate terms.

WORK-HOUR FLEXIBILITY

This second form of flexibility directly reflects the power that the principal contracting company has over workers operating on their premises. The owner-customer states its demands and it is up to the subcontractor to manage working conditions and work hour organization in such a way as to meet those demands. Labor legislation rules have been greatly relaxed over the past few years concerning workload organization, but they still run contrary to those the customer wants to impose. The answer to this contradiction is flexible work hours, which are often at the border between legal and illegal and sometimes on the far side of it. Flexibility in this area can take three forms.

Anarchic Schedules

Many employees no longer experience any sort of regularity at all in their daily, weekly, or monthly schedules. Companies not only resort to night work and ask for special dispensations concerning Sunday, the weekly day off; they may also stretch the workday far beyond the number of hours legally permitted and seven-day weeks of work may pile up without any compensation in the way of rest. The results of the STED survey show that over half of outside workers have had this experience (Doniol-Shaw et al., 1995).

For Eric and Fabrice (valve mechanics), the obligation to produce results for EDF makes it impossible to avoid organizing work hours that way:

> Eric: Normally, we're not allowed to work more than ten hours a day.
> Fabrice: But sometimes we do 15 or even 24 hours at a time, but when we punch in we manage to transfer some hours from one day to the other.
> You cheat a little on the hours?
> Fabrice: No, it isn't that—it's that we have something to do for EDF. But given all the stuff they make us do with their specifications . . . it becomes impossible. Finally, what we were supposed to do in three weeks, we've got to do in one. So we set up shifts, and often as not, we double or even triple them.

Denis (nuclear servitude worker) explains how EDF subcontracts the matter of managing unforeseen time constraints to the outside companies, and the repercussions that those factors have on organization of employees' work hours:

> Just last night we were scheduled to do surveillance on the secondary steam generator level. But our foreman was called in (by EDF) yesterday afternoon, at 7 P.M. They told him the surveillance had been cancelled, so in fact he had made us come for nothing. But since we hadn't filled our quota of hours [weekly], they had to make us work anyway so that we would have our weekly hours. So we went, we fiddled around, you know. And tonight it'll be the same thing, since surveillance only starts again on Monday.
>
> What do you do exactly?
>
> We keep busy! There's always something to do. We keep busy. But if we hadn't been here it wouldn't have made any difference. But our foreman only found out yesterday at 7 P.M. It was too late to reach the guys, to tell them "Don't bother coming." So there you have it! (. . .) They could have sent us home but since we didn't have our weekly quota, they absolutely had to keep us on so we'd do them because they'd already given us our Thursday. If they'd known, they wouldn't have given us our Thursday and we wouldn't have worked on the weekend. Now we're going to work all weekend for nothing . . . just to fill our hours quota.

According to the STED survey, for over half of DATR workers, work-hour flexibility is accompanied by irregular mealtimes and sleep. Many studies have shown the pathogenic effects of this social use of work hours, particularly in terms of premature aging. (Teiger, 1981).

Waiting Around and Bureaucratic Formalities

Given the interference between the activities of so many companies during unit shutdowns and the number of outside workers present in the same nuclear power plant simultaneously, anarchic work schedules, together with the waiting periods they entail, are particularly trying on employees' nerves. They know that work remains to be done and that it will have to be done regardless of how much time it takes. In the STED survey, nearly 75% of employees declared they were bothered by the waiting periods. If waiting is done on site, normally it is counted as work and paid. But when workers are sent home to wait, they are on call without compensation.

Two statements—one by a nuclear servitude worker (Patrick) and one by an EDF agent—illustrate what these waiting periods are like and the lack of any rules or regularity in the use made of "outside" maintenance workers' time.

> Last week, I arrived Monday morning at the D. reactor, which is 300 kms away from where I live. We did the formalities at reception on Monday morning. We had to be there at 8 A.M. They had sent for us for a decontamination that was supposed to begin Monday evening at 11 P.M. But it only began on Thursday. We waited for three days in a camping ground without being able to go home. We had to be there in case EDF asked for us. There were some loading problems and also the EDF agents' inability to manage things. It's the present situation that's to blame: decisions are made

> with an eye to saving money. The line is "Do unit shutdowns in record time." We're rushed all the time. Thursday we started at 8 P.M. and we finished Friday at 4 in the morning. We were told at 4 P.M. that we had to start again at noon. We waited four hours in the locker room. Then we started again at 8 P.M. and kept going until 8 A.M. on Saturday.
>
> We worked non-stop for 26 hours in controlled zones. They give you short breaks. But on Monday, for example, we left Tours [a French city in the center of France] at 6 A.M. We began at 9 A.M. Then we did 26 hours. I started shouting, saying I didn't want to work under such conditions. I like to move around, but not to do just any stupid thing.
>
> I've been decontaminating tanks for two months. That's supposed to be a promotion. It's horrible, it's awful. It's "Hey you there: work! Hey you there: eat! Hey you there: sleep!" (Patrick)

An EDF agent in charge of unit shutdowns explains how part of the EDF hierarchy uses its position of power to impose this kind of time limits:

> The guys who work with us, the company that hires them, they decontaminate tanks. They're guys who work with us eight hours during the day, but they can be called back in the evening to do something else. Since their company has several sites going at the same time, as soon as they need personnel. . . . Well, those guys, you see them again the next morning and they're worn out.
>
> A technical manager asked for some private-sector agents for scaffolding, since they're the only ones available on Saturday. It was the time kids were doing their communions. The team boss of that company couldn't fill the order for Saturday, and to top it off, I think it was already Thursday. He only had two or three people who could come do the work. The technical manager said: "It's not my problem anyway, we've got to have at least ten people, so do whatever you like, but do it." That made a few waves. It ricocheted back by other people saying that they're not animals but employees like everyone else. But it shows exactly what the superiors think of subcontracted service companies—or what part of them think. Their guys are here to work, period. They can just keep their mouths shut.

Some plants require outside workers scheduled to begin work on Monday morning to check in already on Friday. When the power plant is far from their place of residence, they have the "choice" of spending the weekend alone or going home in spite of the distance.

> We used to do the formalities at reception on Monday. You went to reception, got your papers, went to the infirmary for the anthropo-gammametry tests, took the tests, etc. It took all Monday morning. Then in the afternoon you looked for a room or a camp ground. We had Monday off. EDF paid the guys for Monday. Now they no longer pay. You get there on Friday morning, even if it's 700 kms away like F. . . . The guys finished their job at the G. power plant on Thursday. . . . They've traveled 700 kms, checked in on Friday morning at 7 sharp. In some plants, that's the way it is. If it's

> 7:05, they close the door in your face. So the guys check in. It takes them two hours, depending on the number of people there. Then what do they do? They go home. That's 1400 kms in a single day and they've got to be back on Monday morning.

One consequence of the requirement to travel by night is road accidents, as Christian, a nuclear servitude worker, explains:

> One of my pals died. He had been working at the Chinon reactor and was supposed to be at Tricastin the next day—that's 600 kms away. He made a stop at his parents' that evening, then took off directly for the Tricastin reactor. He fell asleep at the wheel, ran into a truck. . . . He was 20.

Intensified Work

The third form of work-hour flexibility imposed on the employee is time limits for getting the job done. As we saw in previous chapters, the reason for the forced time limitation is the intention to decrease the amount of dose received. It is also meant to shorten total operation time.

Jean-Marie, a skilled construction worker, began to work on reactors in 1987 as a temporary RP2 and worked until 1990, when he was hired by an electric company first on fixed-term then on unlimited contracts. He accumulated experience in several nuclear plants, first as a regular service worker then on a package deal. In July 1993, he was laid off without notice after two warnings for insufficient productivity. At the time of this interview, he was once again working on a nuclear site, once again as a temp.

He explains that the problems with his employers started when they asked him to

> tighten up the timing, thus making the intervals shorter and shorter, whereas to do something well, you've got to take your time. But when you're in a package deal, in two hours, the job's got to be done, even the paperwork on the computer, the reports you've got to type out, everything. It doesn't make sense. There are too many requirements.

Jean-Marie received two warnings because he had fallen behind.

> A job that had to be done in 30 hours took me 35. And since I couldn't justify those five extra hours, I got warnings.
>
> The man in charge came to see me and said, "Don't worry your head about it, just put on fewer clamps." You used to see cable holders put up with the cables in neat piles. You see fewer and fewer of those today. It's even scary. On one hand, they make you take QSP training for security and on the other hand they tighten up the job time. If you think about it, it's impossible. And I'm not the first to be laid off for professional misconduct.
>
> How do they explain that it's professional misconduct?

> Well, they said I was too slow. But it's not easy to do a good job with somebody watching you and telling you to go faster. And it was connected to a job that had been badly done. It worked allright but it wasn't aesthetic. They didn't like the way the cabling looked. But it worked, and the colors were right according to the plan. What's more, I'd never done a cabling like that one before. So I made a mistake. But firing me for that is really disgusting.

Jean-Marie took his case to the industrial tribunal and got 50,000 francs compensation [around €7600], but he's once again working as a temp.

François, a 29-year-old "veteran" inspector who has been working in nuclear industry maintenance for seven years thinks that such an incident makes it extremely difficult to use what one has learned during the training periods. Also, "the lifestyle" imposed by the organization of labor in the nuclear power plant is ruining his health and has caused him chronic intestinal problems:

> We're the ones who're going to pay for the next generation. It's hard. If the system doesn't get better now. . . . There are holes in the way things have been thought out. There are texts, requests for training of service providers, but it's not always applied in reality. I'm not saying it's not applicable, but it's not always easy.
>
> The future? I'd really like to continue, but I'd like to have a broader perspective and vision, try to improve things. But it can't be done starting from nothing. The company has to have the wherewithal, it's got to make it a possibility and you need competent people. And then nothing proves that EDF will buy what's being offered. They're in command, after all. And in the company, on the training front, not everything's perfect. The company is only interested in productivity. What it sees is whether you do your job well, if you sold yourself at a good price and are making profits for them. And they're constantly speeding up the production rate. The boss wants to spend as little as possible and put in as little time as possible. I don't mean we're loaded down with work 24 hours a day, but there are work paces that are really difficult, stressful. You've got to be available all the time. If you wanted to apply literally everything they ask you to do and do it well at the same time, it's impossible. They just don't give you enough time. They carried out research on this and noticed that such and such an operation which takes ten hours could be done in four, and time is being reduced more and more. It's getting to be a hellish race.
>
> And your intestinal problems are connected to that?
>
> They are. I wouldn't say it's because of a bad lifestyle, but it's linked. You try to make your life as comfortable as you can given the means available, but it's not easy.

Like an echo to this, an EDF agent criticizes the time pressure put on outside company employees:

> The [outside company] boss is playing a game. He's blackmailing you with the job, it's not even real employment anymore. Just short 2-week or 3-week, or one, two-month contracts.
>
> When they're taken on they're hired for three months, they're cooks working as mechanics. I don't want to be nasty, but that's what it is. Those guys put in twelve, fifteen hours a day.
>
> I've noticed that some persons from the outside companies even sleep in the locker-rooms of the nuclear power plant because they've got heavy work schedules, their jobs are all broken up, four hours in the morning, eight hours in the afternoon, and they come back at night. They sleep and eat in the locker-rooms.
>
> Shortening unit shutdowns means they have to work a lot faster, even less safely—and there's already not a lot of safety—and sometimes they bungle their work or don't even do it at all. It does happen that people don't do the work that's required.
>
> When they're called in on Friday–and I insist on that last day of the week—when they're asked to do a job and that job has to be finished at noon on Friday or at least by the end of the afternoon, the guy has other things on his mind, so if there are safety measures to be taken, more often than not he'll overlook them in order to finish the job, to be able to leave on time.

Job and work hour flexibility, as well as the fact that the employee has to be constantly available in terms of both time and space, go together with forced flexibility of one's private life.

FLEXIBILITY IN FAMILY LIFE

The availability required of subcontractor company employees works to maintain and reinforce the traditional sexual division of labor in the home, making the woman entirely responsible not only for the organization of family life but also for the effects on her own limited choices of the fact that the man's work puts him at the beck and call of the company.

There are several sides to flexibility in family life: the effects of seasonal and intermittent work; the effects of permanent availability and geographic mobility, and individual and family upheavals due to the lack of regular daily and weekly routines. After a short presentation of these various aspects pertaining to family flexibility, they will be illustrated through the testimony of the widow of a valve mechanic who died of bone cancer in 1992.

The Effects of Seasonal and Intermittent Work

The results of the STED survey (Doniol-Shaw et al., 1995) show that maintenance workers in the nuclear industry are increasingly "intermittent," regardless of whether they are employed permanently or temporarily. The average yearly workload is 10.3 months for employees with long-term contracts, 8 months for those on short-term contracts, and 6.8 months for temps.

This means that periods of intense professional activity are followed by forced inactivity, making the organization of family life quite difficult. The first period corresponds to the summer months, meaning that vacations as a family become impossible during the main school summer holiday. This also means that the woman takes on the entire responsibility for the children during those periods.

On the other hand, during periods of inactivity, men who stay at home are not only cut off from work but also have no legitimate social excuse to be there, since the wives have usually taken on all the responsibility of organizing family life.

This imbalance creates problems in couples, problems which were mentioned during the interviews, discreetly expressed in accounts of the tremendous difficulty respondents have managing the permanent contradiction between their occupational and private lives.

The Consequences of Having to Be Available and Geographically Mobile

Maintenance workers must be ready to get up and go at all times, from home to one or another nuclear plant or else between two plants, often without any prior warning. For the men's wives and companions, this means a difficult choice (or the absence of any possibility to choose): either accompanying their partner in his movements and therefore accepting precarious living conditions and renouncing any sort of lasting, personal, professional or social integration, or staying put and therefore living in a place where the man will only be passing through. The way men and women expressed the suffering connected to the need to make such a decision illustrates how little leeway they have at a time when unemployment is so high that in fact there is no alternative. The film "Nuclear hobos" ("*Les trimardeurs du nucléaire*" by Catherine Pozzo di Borgo, 1996) contains particularly eloquent testimony on this.

Individual and Family Disruptions Caused by the Absence of Any Regular Daily or Weekly Routine

Women talk about the problems typical of each of the above situations. When they accompany their husband, they live the hardships of camp life, and have to swallow being looked down upon by others as "hobos." Instability obliges them to be constantly reorganizing the details of daily life and the best ways to help the children adjust. With nothing to do on a campsite, which can be far from everything and in some cases unwelcoming, particularly in winter, the solitude and isolation are often hard to bear.

Solitude is also the lot of couples who opted for the other solution: a house that the man must leave to shuttle between plants, while the woman remains alone without being able to count on regular homecomings and the presence of the man in a routine that would allow them to structure the moments of family life and share child-raising and household tasks.

Solitude is compounded by anxiety due to night driving, often for long distances, but also due to the very nature of the work and exposure to ionizing radiation.

The Right to a Family Life?

The social organization of subcontracted maintenance activities in nuclear plants challenges the right to family life. The narrative of the wife of one of these workers illustrates this.

Céline has three children. When her husband died in 1992, they were 10, 6, and 2. He was 39 when he died. Céline works part-time as a service technician in a rehabilitation center for senior citizens.

> He was an electrical mechanic. He left school at 14. He learned on the job. He got his DATR in 1981. He did the plants: Bugey, Fessenheim, Tihange in Belgium, Flamanville, Porcheville, Dampierre, Chinon. . . . When it was less than 400 kms away, he came home every week. He stayed in rooming houses. When he worked in Cattenom, he went there with our mobile home.
>
> He didn't like to talk about his work too much . . .
>
> The last site he worked on was in September in Belgium. He hurt his eye at the plant while he was welding. He came home Saturday evening because Saturday morning he had to work. Sunday morning he went to the hospital because his eye really hurt. The doctors wondered if his pupil hadn't been damaged. His doctor wanted to put him on sick leave but he came back with a patch on his eye, he didn't want to leave the job. . . .
>
> You know, when I saw him leave on Sunday night, knowing he had to drive 600 kms and that he had to be at work the next day, I was sick with worry. On top of everything, his foreman phoned me when he didn't arrive. . . . One Sunday evening, he had left at 11 P.M. with a co-worker. It was snowing. In the morning, at I can't remember what time, his boss called: "Your husband isn't on the site." "But they left last night," I said. "We'll call you.". . . At noon still no phone call. At 4, still nothing. I wondered what had happened. Finally I called myself. His boss said "Don't worry, if something had happened, we'd let you know." . . .
>
> He had an accident, one Friday evening on his way home from Fessenheim. He had tests at the hospital and they let him go on sick leave for a month. . . .
>
> The times he got dosed—one month he took in 2300 millirems [above the yearly allowance according to the new regulations]—when he came home he'd sleep all day long and I had the feeling he wasn't reacting. Well, he reacted in a way, but he wasn't really there. And his friend was the same. . . . When he worked at Bugey I was more frightened than ever. I really had the impression he wasn't registering what I was saying. After that, they were forbidden to enter the zone for several months, from November 1987 to April 1988. . . .

His sickness began with a neck ache. He went for a DATR medical visit on October 21st. The company doctor told him it couldn't be fatigue from work because he wasn't working. He was waiting for a site. But it didn't go away, and aside from that he started having stomach pains. Since it was getting worse, I called our family doctor. He said he should have a check-up and sent him to the hospital on November 14th. When he left, his arm had gone into paralysis, he could practically not use his hand. He had a scan on December 5th and they saw he had lesions at the cervical level. The next day, they called me, they told me it was serious, that he shouldn't lift the children at all and that he was going to be sent home lying down in an ambulance. In December, the surgeon operated on him and saw that the tumor had done a lot of damage. He unblocked the nerve and sewed him up. That was on December 24th. . . .

They're not the ones who told me, only my family doctor did. With my husband it was the same, I could never speak with the doctor because he would have known and he would never have allowed me to ask about his problem in front of him. He would have said, "It's my illness, it's my business, it's got nothing to do with you." They decided not to tell him because he was pretty compulsive. There were times I wanted to tell him. But the doctor had told me not to tell him. But it's true there were times I wanted to tell him. Because I saw him suffering so much and besides, he was making plans for the future whereas I knew for sure that. . . .

He wanted to change jobs. The doctors had told him that whatever happened, he wouldn't be able to do the same job as before. He knew it. But it was all the same to him, as long as he could work.

He was hospitalized for the first time on November 14th and he died on April 10th. . . .

The specialist at the hospital thought it could be connected to his work in nuclear and he asked us if we would agree to apply to register his condition as an occupational disease. . . .

Did your husband think there was a connection?

I don't know.

Did he talk about the nuclear danger of his work?

Yes. Several times he had to stay up there over the weekend because there were people who had been contaminated. He spoke to me about the risk of becoming sterile because it's true we had trouble having our second child. And it's true I was scared. But he didn't talk to me about it. . . . He knew I was very nervous. . . . I had convinced myself that if he was contaminated, he could contaminate us too. . . . When he went into a zone, I was afraid, I was afraid he'd be irradiated, so he preferred not to talk about it. . . .

The hardest part is that today he's gone. The kids never had him to themselves. . . . He didn't see them grow up. There was only February, when he was sick. . . . I remember once he talked with our oldest son until 4 in the morning, he'd never done that before. Either he wasn't here at all or else he was awfully tired. And I have the feeling that that night, when he talked with our son, he wanted to make up for lost time. He asked him what he wanted to do later on, and he told him about his own youth. . . . But first, well, he asked how the kids were getting along, how they were doing in school. . . . Once we went for a vacation in Poland, it's the only time. And once we took the mobile home to C. . . . I had asked for my vacation time but he had to work and instead of staying here, we went to spend a month up there with him, that way he could be with us every day. . . .

He really liked his job. I know that for a while he thought about working on highways and had looked into it. He was of two minds about it though: it would have been nice because he could have come home every night, but it was a six-month contract and they didn't know if they'd keep him on once it was over. . . . On the other hand, I knew he'd left school at 14, and it had really been hard for him to get to where he was. Was he supposed to leave all that for who knows what? . . . And after all, he really did like his job. With his friends, they talked shop all the time. Sometimes I told them, "Stop already!" But it was their life. It was their life on the site. They sure spent more time there than at home. . . .

It's true since a year ago he'd wanted to stop moving around. He realized the children were growing up and he was missing out on a lot of things. Towards the end, it got harder and harder for him to leave. When the children were small, it was different. But then he saw that, well, our oldest boy is going on eleven. . . . At that age, a boy needs his father more. . . . They went fishing together, they did things together. . . . And it was also tough to be alone all week. I said, "for me too." "Yes, but you've got the kids," he said. It was true. But he had his pals on the site. Come evening, they all ate together, and after that he was all alone. . . . It's hard. And also, you become independent. It's true that we had problems as a couple because I became independent too. I had to, because of his work. The kids and I had our routine during the week. We ate early, whereas on Saturday he liked to be with his friends. We both became independent and there were times we couldn't stand each other anymore. When he was home for a week, it's true it got to be pretty heavy, for him like for me, he knew that too. He once said "All those first years when we should have been together, we lived like unmarried people." . . .

My children couldn't understand. I could explain until I was blue in the face that it was for his job. But they had friends whose dads came home every night and they couldn't understand why their father didn't. And since we lived in a project, they saw all the other dads come home at night. For a while, the oldest was afraid because he had a friend who had never known his father. That friend told him once, "My dad left home one day and never came back." My son started getting obsessed. He didn't sleep nights. I told my husband, "Try to phone him and talk to him." Well, he wasn't the sort to call home every night, but that's not why he didn't do it. I once talked it over with his friend. He said to me, "You know, he's always talking about the kids." But it wasn't his way to phone in the evening to see how things were. And when he left the last time for a whole month, it's true the kid was sick. He had stomach aches. He stopped eating. It was the first time he was gone for a month. . . .

The procedure for getting this worker's physical condition recognized as an occupational disease failed. That issue will be taken up in the Part II.

CONCLUSION

Labor law regulations are aimed at loosening the ties that bind and subordinate employees to their employers. However, Céline's narrative and the considerable amounts of testimony presented in this first part show that regardless of whether we examine work hours, the conditions of exposure to radiation risk, intensified and accelerated production rates, or what right workers have to withdraw from dangerous situations, these regulations have somehow become obsolete.

The qualitative results presented in this first section are validated by the data produced by two statistical surveys on working and living conditions of employees of subcontracted companies in the controlled zones of nuclear plants. Our results fit with those of the STED survey (Doniol-Shaw et al., 1995) conducted in 1993, i.e., at the time the "outside" workers whose accounts are presented in this book were being followed up. Another statistical survey was carried out in 1997 at the behest of the authorities in charge of France's nuclear industry, by the *Centre de Recherche en Gestion* (research management center) of the Ecole Polytechnique in Paris (Dégot, Globokar, & Girin, 1999). This study was based on 90 qualitative interviews with managers and workers of subcontracting companies and 1368 questionnaires distributed to outside employees operating in four different nuclear power plants.

The results of the CRG study converge with those presented in these first three chapters concerning four dimensions of the situation. The first is the disparity between what is said and what is done in matters of organizing unit shutdown work. With regard to "the prescribed," employees of subcontracting companies acknowledge that much attention is given by maintenance managers at both EDF and their own companies to training and task specifications. On the other hand, in matters of "practice," over half of the people questioned in 1997 feel that nothing has changed in site organization practices, which remain "the equivalent of rush jobs and are characterized by overlapping activities, crowding in places where operations are underway and frequent delays." The second point concerns the problems of managing dosimetry. These problems persist. It is true that the number of DATR agents who exceed the yearly mSv allowance is diminishing, but this improvement is obtained mainly through managing jobs assignments by dose, a point raised—though not explicitly—by the Ecole Polytechnique researchers. They also note: "The ALARA philosophy, which tends to divide up a given overall dose among all participating workers, runs up against the social stratification that cuts across the worker population, especially the division between technical, 'noble' skills and those which are more strictly manual and thus less noble. For some members of the 'noble' category, it would be better for the less skilled workers to receive an overall higher radiation dose—in setting up protection systems, for example—so that highly specialized workers, more costly and harder to come by, will be less exposed." This is of course the idea of social division of radiation dose (see chapter 2), a phenomenon sufficiently legitimate to be openly expressed in the framework of a survey study. Third, the CRG report points up the triple flexibility—jobs, work hours, and family life—and the "stress" (in the ordinary sense of the term) experienced as the result of it by outside company employees operating in nuclear plants, as well as probable violations of labor law, particularly in terms of work hours. The fourth point on which CRG study results and my own converge is denying workers recognition of their skills and not allowing them to participate at all in planning maintenance operations or in "return on experience" assessments of those operations. This situation is considered "worrisome" by the Ecole polytechnique researchers: over 80% of subcontractor company employees interviewed wanted to stop working in the nuclear industry.

In a public-sector company such as EDF whose employees are particularly well protected in terms of social welfare, job security, and wages, and where parity between management and labor unions has turned negotiation into a permanent feature of the industrial landscape, how can such a change in employee working conditions—involving a large segment of nuclear power industry labor—be accounted for? That is the question to be tackled in Part II, which focuses on the role played by strategies for managing the risk of exposure to ionizing radiation.

PART II

Sources of Legitimacy: Competitiveness, Safety, Radiation Protection

The social organization of maintenance work in nuclear plants results from a combination of various types of logic based on processes of legitimation that allow EDF as a company to sideskirt the contradictions inherent in nuclear production, contradictions between keeping the nuclear plant functioning and keeping it reliable, as well as between safety for the population at large and on-site radiation protection.

The first type of logic, competitiveness, dominates the present-day industrial world. In the logic of competition, all attempts to reduce costs are legitimate in and of themselves. Labor costs are at the core of this undertaking. Reducing costs is nonetheless very difficult because of the labor agreements that have been in effect at EDF since 1945. The status of EDF personnel as determined by labor negotiations between management and the unions includes highly specific guarantees regarding qualifications and wages, working hours, working conditions, and vacations. These guarantees cannot be called into question without new negotiations. Subcontracting has allowed EDF to break with the management-union negotiation approach in the area of maintenance. This development has been legitimated not only by means of the preponderant value attributed to the inherent competitiveness of nuclear energy—there is consensus in France on this point—but also by legislative and regulatory texts that legalize the division of labor instated by subcontracting.

Subcontracting tasks that used to be performed by the owner company has externalized responsibility for realization and performance of these tasks. This explains why the owner has to check the quality of the work done, a requirement that is part of the general industrial development designated "standardization." Standardization means defining technological quality standards to which companies seeking to acquire a quality label in a given specialty must then conform. "Quality assurance" has gradually come to include not only safety standards but also behavioral norms for maintenance operators. The approach adopted by EDF in the areas of quality, general safety, and worker safety, and the associated requirements for both EDF agents and subcontracting company employees are to be understood in connection with this development.

But the specificity of the nuclear industry presupposes particular, on-site protection of workers exposed to ionizing radiation. Given France's high number of nuclear installations and their age, managing exposure to ionizing radiation is a major problem. EDF has chosen to resolve this problem by side-skirting the regulations adopted in France with regard to radiation protection as well as occupational health legislation, instituting instead a system of managing employment by radiation dose. This system has been legitimated by the Commission Nationale de l'Informatique et des Libertés.[1]

Chapters 4 and 5 of Part II examine the process of legitimating the particular social division of labor and the associated risks brought to light accounts from subcontracting companies employees presented in Part I.

[1]National Commission on the Computerization and Freedom.

CHAPTER 4

Competitiveness, Quality, Safety

In the production system of the 1990s, instrumental rationality was used to attribute a single, ultimate value to changes in labor organization and the meaning of those changes. That value is competitiveness (Petrella, 1996). Competitiveness imposed itself as "natural" and primary, invalidating all reference to any other values that might pertain either to the economic production of energy or the social forms of such production. Competitiveness goes a long way to explaining the strong social consensus around France's choice of nuclear energy. This legitimacy is summed up in a single slogan: "Nuclear energy or candles." The search for ways of cutting production costs so as to enable EDF to maintain the competitiveness of nuclear energy as the energy of the future is therefore crucial to the enterprise as a whole. In 1994 an agreement was signed between the French state and its company EDF requiring the latter to reduce the price per kilowatt/hour by 1.5% a year.

In fact, from the nuclear industry's beginnings and particularly after the accidents at Three Mile Island in the United States (1978) and Chernobyl in Ukraine (1986), competitiveness has been thought of as indissociable from safety, i.e., the set of arrangements for preventing accidents. The quality of maintenance operations is of course a primary focus here, as are procedures for verifying that quality.

GETTING TO THE SOURCES OF PRODUCTIVITY AND COMPETITIVENESS

As is the case in all integrated, automated technical processes, nuclear industry productivity is linked not only to the production process itself but to two determinant factors brought to light by Philippe Zarifian in his analysis of "new productivity" (Zarifian, 1990).

The first is "reducing the amount of time that machines are out of operation." For the nuclear industry this refers in particular to unit shutdown time, during which fuel is unloaded and nuclear activity stopped. During this period the nuclear plant is of course significantly less capable of producing electricity. This is a major preoccupation for those in charge of France's nuclear installations as a whole. In the 1980s, for example, plant "capacity" (operating availability) decreased from 85% (1986) to 71% (1992). This led to the decision to shorten the period for maintenance operations, which as we know is now limited to April-September, as well as individual unit shutdown time. By 1994, capacity had gone back up to 82%, a rate that seems relatively stable (82.7% in 1997).

The second factor is "reliability of installations and processes (reducing chance events and breakdowns)." In the nuclear industry, this means not only guaranteeing plant operation (and therefore productivity) but also eliminating all danger of accident or radiation release into the environment. It is this requirement that is termed nuclear safety. But nuclear plant aging increases both wear and tear on installations and radioactive contamination. As we have seen, contamination affects maintenance workers' degree and conditions of exposure to ionizing radiation.

To guarantee nuclear installation reliability, it is important not only to be attentive and proactive with regard to operating and aging processes, but also to anticipate the consequences of wear and tear. It may not be known what form such wear and tear will take or the timeframe it will occur in, and though events or developments such as cracking in fuel cladding, cracks in piping or tank covers, leakage of radioactive effluents, lead corrosion of piping, valve defects, welding that is no longer leakproof, etc. may be either likely or unexpected, they are in any case unpredictable. They are what are called in the nuclear industry "fortuitous occurance" of unplanned events.

Properly maintaining nuclear plants thus presupposes permanent maintenance work in which anticipation is crucial for limiting the probability of breakdown, failure or accident. One crucial element of that activity is interpreting signs of aging, threats of cracking, wear on welding or seals, defects in metallic structures, and wear on the material components of the nuclear cycle technical system. Maintenance, then, involves unwavering surveillance of signs and varieties of wear on materials as well as surveillance of the multiple mechanical, electrical, and electronic systems that work together to produce nuclear energy.

Though help in interpreting such signs of wear is to be had from technical and electronic systems, the tasks of decision-making and implementing the necessary corrective measures cannot be delegated to robots. It is imperative to "go see," and to work in places where the threats are greatest with regard to both installation safety and safety of persons performing the tasks; namely, the reactor building, the primary circuit, and the steam generators. These areas are always dangerous because always radioactive.

This is one of the major contradictions that nuclear industry managers are confronted with: nuclear safety depends on the quality of maintenance operations, operations which depend absolutely on human beings; but the older a plant is, the more dangerous it is with regard to irradiation and contamination, and this of course increases degree of exposure for persons in charge of these operations.

It is important here to raise a question about the preexisting situation. Who was in a better position than EDF agents to ensure the proper organization of nuclear power production? Permanently present on plant sites, endowed with long-term experience of nuclear plant operation and incidents, these workers represent (or represented) the real memory of the plants, which are installations relatively new to industrial experience. Given that the electrical industry in France is a state monopoly originally conceived above all as a public service and not an activity subject to international competition, why didn't EDF and the state choose to make use of their experience and know-how, their "memory" of nuclear power installations, to ensure the safety of both on-site workers and the neighboring populations? Why did they choose to subcontract nuclear industry maintenance work?

To answer this question requires analyzing what subcontracting represents for production organization as a whole, after which we will consider the particularities of the nuclear industry in this connection.

Subcontracting

Altersohn, a specialist of subcontracting at the French ministry of industry defines subcontracting thus: "a technique that allows an economic agent to unload onto another tasks that it is legally responsible for because they are obligations agreed to in the framework of its own activity" (Altersohn, 1990, p. 16). The author thus points out "the existence of relations of domination linked to the very nature of the subcontracting relation, a relation created through often random choices between 'doing' or 'having done by someone else'" (Altersohn, 1990, p. 19).

Through what modes are these relations of domination actually exercised? In a study of *"subcontracting and labor relations"* from a legal standpoint in two very different industrial sectors, Morin (1994) shows that "asymmetry in relations between principals and subcontractors" may be observed at two levels.[1]

The first concerns the economic relation whereby principals pit subcontractors against each other, making them compete in continually renewed bidding for jobs where absolute priority is given to an approach in terms of *"lowest offer."* To prevent degradation of the necessary quality of operations, however, this purely market-oriented logic is today contained by a preselection instrument; namely, the fact that any company wishing to compete in bidding has to adopt the *"quality*

[1]"*Principal*" is the term used in this book to translate the French expression "*donneur d'ordre*," the entrepreneur which outsources work to subcontracted companies (subcontractors).

policy" and acquire AFNOR certification (ISO 9000). The matter of quality standards will be examined further on.

The second concerns the diverse forms of *"technical subjection"* that require subcontractors to satisfy principal's demands in matters of technical procedures, quality, deadlines, and "risk-sharing." In other words, principals do not hold themselves responsible for the hiring problems that their subcontractors may encounter due to fluctuations in principal's own activity. Recourse to temporary workers or work-force loans become the only possible strategy for subcontractor companies faced with these fluctuations, the only way not to imperil their economic survival and competitiveness against other subcontractors.

Morin has shown how wage level differentiation and difference in collective accords play an important role in increased recourse to subcontracting: they allow principals to reduce labor costs without having to negotiate those decreases with labor unions in their own companies. In formal terms, wage conditions for principal's in-house employees are not threatened. As for wage conditions of subcontractor employees, that is an issue for those companies alone, and action by their own representatives in the companies that employ them. In a study of the change in relations between hiring and employment processes and labor law brought about by company restructurations, Rettenbach (1978) underscores the way contractual relations between work prescriber and work performer have been dismantled by subcontract work. The dismantling of these relations also has the effect of dividing up and reducing the space in which union representatives of the various companies are officially able to act, despite the fact that these companies' permanent employees participate in one and the same production process, organized by the principal.

Meanwhile subcontractor employers are caught in two contradictions. First, they are forced to meet technical and quality demands that are constantly being stiffened by the principal, who is also increasingly tightening timeframes and seeking to lower costs. The second contradiction is that obtaining between the set of constraints imposed by the principal and the work, hiring, and wage conditions of subcontractor employees.

Morin writes that labor law has no means of impacting on the relations of domination inherent in the subcontract relation. Those relations obtain not between an employer and employees in the framework of a contractual relation accompanied by guarantees, but rather with subcontractor employees who are completely subject to the competition game in the framework of a "customer-supplier" relationship between the principal and their employers.

As we know, the foundation of labor law is the work contract. Suppiot (1994) in *Critique du droit du travail* points out that the "primary *raison d'être*" of labor law "was and continues to be to bring employees into being as rights-bearing subjects within the company; that is, to 'civilize' management by endowing it with a legal framework in which to operate." It is precisely this framework that is marginalized in subcontracting relations, because the power-holder—i.e., the principal—does not use that power with reference to rights and duties specified in a work contract but through a commercial contract between companies, one that has no "labor provisions" for employees' hiring and work conditions. In principal company accounting, subcontracted work disappears from "human resources" and falls into the "purchases" column.

Subcontracting a number of functions (including productive ones) is thus a kind of social organization of labor that allows principals to artificially diminish payroll (and therefore official labor costs) and throw off the yoke of stable employee wage relations. Instead they rely on and use the flexibility and power relations characteristic of customer-supplier relations. This allows those companies to transfer the financial and organizational burden and responsibility of managing non-work periods, training, safety, and social protection (unemployment, health, and retirement insurance) to other actors. It also allows them to delegate the matter of implementing their human resources and organizational demands and imperatives in matters of hiring and working conditions, time constraints, and productivity to other companies without having to deal with either the financial or social consequences of this; namely, without having to negotiate these imperatives in the framework of representative labor institutions.

Subcontracting maintenance work—work that involves a high level of unpredictability and fortuitous occurrences—allows the principal to transfer responsibility for physical, on-the-spot management of such occurrences to those who have to perform the work. In all sectors of industrial activity, a particular division of labor has been instated between principals and subcontractors for maintenance activities: principals do the conception, preparation, oversight and verification work, while subcontractors perform the operations in the framework of a strict, precise set of written specifications. The customer-supplier business relation excludes any form of contracting between the principal and employees involved in this activity. On the other hand, all demands from the *"customer"*—even if they contradict the *Code du Travail*—become imperative, since customer satisfaction has become the guiding principle of the activity. In case of a breach of the *Code du Travail,* it is not the principal who is accountable, but the subcontracted employer whose employees are understood to have committed the breach. Indeed, the work contract at the base of labor law prescriptions is established independently of the principal, between subcontractors and their employees.

The only text in the *Code du Travail* that mentions principal's responsibility toward subcontractor's employees is the decree of February 20, 1992, concerning "particular prescriptions for hygiene and safety applicable to work done in an enterprise by an outside company." This decree reads as follows:

> The director of the user company [principal] ensures coordination of prevention measures used by that company and prevention measures taken by all managers operating in his establishment. Every company manager is responsible for applying the prevention measures necessary to protecting his

> personnel. . . . The purpose of this general coordination is to minimize the dangers linked to interference between the various activities, installations and materials of the various companies present in the same workplace. . . . To further such coordination, the manager of the user company is called upon to alert the manager of the outside company when he knows of a serious danger concerning one of that company's employees, even if he judges that the danger is due explicitly to that company, so that prevention measures can be taken by the employer(s) concerned.

This text does not recognize full responsibility of the principal vis-à-vis all workers who come onto its site. Rather it defines the job of coordinating prevention measures, a job the principal is in charge of, referring quite explicitly to the subcontracting company's obligation to maintain worker safety. The text seems aimed not at establishing principal's responsibility for risk exposure and work conditions for employees of companies operating on its sites but rather at freeing the principal from that responsibility by making subcontractor employers responsible for applying prevention measures, including those related to risks linked to interference among different on-site activities. The principal's CHS-CT prerogatives are therefore very limited with respect to working conditions for outside company employees. However, the CHS-CT may always legitimately intervene in any dangerous situation involving "outside" workers, though this is only possible if it is aware of those situations. Awareness is itself a function of the social organization of work on site and whether or not principal's employees can meet with employees from outside companies.

Maintenance in the Nuclear Industry: Doing or Having Done by Others

Starting in the 1980s, the choice between doing nuclear industry maintenance or having it done was made in favor of the latter. This choice, consistent with changes in the social organization of industrial work underway since the mid-1970s, answered to the new demand for flexibility, itself consistent with the new modes of production management and the strategy for loosening the constraint of labor negotiations with unions, a constraint that was particularly strong at EDF given the origins and history of the company (Duclos & Mauchamp, 1994; Gérôme, 1989; Groupe de recherches historiques de la centrale de Chévirẽ, 1987; Janiaud, 1990).

EDF's entry into international competition and its strategy for exporting both electricity and nuclear plants made it necessary to break with the logic of labor negotiations, which would too tightly constrain policy on personnel management, wages, and work conditions. It was in reference to a discourse on "modernization," the "company project," and new forms of "management" (the different ideological aspects of this approach are presented in Meynaud, 1996) that

EDF management undertook the structural reforms that would lead to nearly complete outsourcing of maintenance.

As in the case of other big nationalized companies, taking the subcontracting path was, for management at EDF and for the state—which did not have to justify it in terms of management rationalization—a way of disengaging from the constraints imposed by the particular status of employees in French state companies such as EDF and from the mode of labor relations operative in that company since 1945.

In 1988, the strategy for managing France's nuclear facilities changed. The idea was now to subcontract almost all maintenance operations; only preparation for and oversight of these operations was to remain internal to EDF. This strategy made use of the existing industrial partnership between the principal EDF and nuclear plant constructors (particularly Framatome and Alsthom), a partnership that had organized the highly specialized maintenance work done by these constructors since nuclear plants first began operating in France. In the five years between 1988 and 1993, the volume of subcontracted maintenance went from 20 to 80%.

The 1990 Rapport Noc

There was no official announcement of the decision to use subcontractors but it was confirmed—without consultation of union organizations—by a report from EDF's thermal energy production department on "improving safety at operating nuclear plants" (Rapport Noc, 1990). The stated objective of the report was to answer the question "How can significant progress be made in matters of quality and safe plant operation with regard to maintenance?" It followed an alarming report by the safety authority on a series of incidents that had occurred in the course of maintenance operations during unit shutdowns in 1989, incidents that the SCSIN (now the DSIN: Direction de la Sûreté des Installations Nucléaires at the industry ministry) ranked as threatening to nuclear safety.

EDF responded to the problem using a technocratic approach to malfunctioning, emphasizing the need to promote a "professional culture," "good, full employment" for EDF agents, and separation of preparation and oversight-verification from execution, as had been recommended by the August 10, 1984 order with regard to quality. This order, which will be discussed in detail further on, did not require subcontracting of execution tasks but did permit it. EDF's objectives were to be realized through two changes:

- developing activities of preparation, oversight and verification, and analysis of maintenance operations within EDF—jobs to be assigned to "permanent personnel." This required changing EDF agents' qualifications and hiring degree-holders only; i.e., with a minimum two-year higher education degree. The aim was a "cultural change" to "functional multi-competence" and development of the "methods and oversight" component of maintenance;

- subcontracting task execution so as to rationally manage "good, full employment":

> For obvious economic reasons, a significant part of these activities [i.e., moving nuclear fuel; disposing of solid wastes; site assistance, including laundering, decontamination, cleaning, assembly, scaffolding, airlocks, handling] should be subcontracted, while control of these operations will be ensured by preparation and technical oversight of them.

The "good, full employment" in question was of course that of EDF agents; this was in fact a way of deflecting potential union protest within EDF. As for the "obvious economic reasons," they are not explicitly identified, but the "occupations" mentioned make it clear that they amount to low-skilled labor that would cost much more if done inside the company than subcontracted outside it.

This development went together with overall reorganization of the structures and forms for managing France's nuclear installations and program as a whole. A permanent structure for unit shutdowns was established. The emphasis was now to be on the EDF agent's accountability for budget use, strict planning of maintenance operations and close scrutiny and analysis of service company operation reports. These reports correspond to the "return on experience" mentioned by outside workers in the interviews presented in Part I.

Distancing EDF "monitoring and verification personnel" from the practical content of maintenance operations in this way was legitimated by giving priority to training:

> Technical control of subcontracted maintenance operations must be acquired through careful, targeted training, rather than by continued experience in performing the operations in question.

Then comes the following astounding affirmation:

> Knowing how to do because one has oneself often done undercuts overseers' self-assurance and authority.

Priority was given to a monitoring process based on written traceability of "returns on experience." The return on experience is not the result of direct communication between EDF monitors and outside company employees with regard to the actual realization of operations, but rather involves information sheets that outside employees fill out according to strictly standardized procedures. It is a matter of transmitting written reports showing how the real work done conforms to the prescribed tasks. Report-writing is an EDF agent's procedure for overseeing the companies performing the operations within the client-supplier relation that links subcontracted companies to the principal. These are not only ties of subordination (different from the employee-employer wage relation) but also and above all relations of domination in which legitimacy falls on the side of the overseer while "outside" workers are called upon to fit the "return on experience" to the standards of EDF legitimacy.

With regard to safety and radiation protection, the Noc Report distinguishes between "operational tasks not implicated in the framework of good, full employment" (namely radiation protection management and verification), which are to be subcontracted, and "functional tasks" (instructions, consulting, training), which EDF agents continue to be in charge of.

A Double Contradiction

The Noc Report thus presents a vast program for rationalizing nuclear maintenance management. But it remains silent on two major contradictions that managers of France's nuclear program and individual site managers alike have to handle. The first has already been mentioned: the opposition between technical safety imperatives and radiation protection constraints. The second involves the increasing demand for nuclear plant capacity, a demand central to competitiveness but in contradiction with the constraints involved in organizing working hours for employees operating during unit shutdowns, since the seasonalizing of those shutdowns brings with it highly uneven yearly distribution of maintenance tasks (concentrated in the April-September shutdown period).

Technical Safety Imperatives and Radiation Protection Constraints

In the discourse of EDF executive managers, the idea of public service has been replaced today by the management ideal of lowering costs. However, EDF is more attached than ever to its image as a company guaranteeing faultless nuclear safety and protection of the population at large from any and all danger of nuclear accidents. The Three Mile Island and Chernobyl accidents put serious dents in that image, tempered by the fact that they did not occur in France.

Safety requires complex preventive and corrective maintenance that allows for anticipating any danger of a breakdown or serious malfunction. Nuclear power plants have to be shut down often so that the reactor and steam generators may be opened, inspected, and treated if need be. But the very operation of the reactor produces radiation and forms radioactive dust that contaminate the valves and pipes. As long as the production system is closed, this radioactivity is held in check, but as soon as it becomes necessary to open the reactor and the different circuits to check their functioning, the radioactivity becomes dangerous. Operations that expose operators to ionizing radiation are therefore absolutely unavoidable.

Clearly, then, nuclear installation safety is at odds with any idea that workers could escape exposure to ionizing radiation. And the contradiction between regulation radiation exposure limits and the high levels of radiation discharge in certain maintenance "hot spots" also has to be dealt with, keeping in mind that certain tasks, such as decontaminating operation zones by rinsing them down, increase radioactive emissions into the environment.

The Noc Report does not mention how important it is not to increase individual and collective radiation doses received by EDF agents. If maintenance work had been kept within the company, what EDF agents would have been called upon to absorb the 80% of collective annual dose currently "subcontracted out"? Here again, it must be noted that by keeping EDF agent exposure constant despite the increase in collective annual dose that has come with the increase in number of working reactors and the aging of existing ones, the executive management of France's nuclear program was protecting itself from strike action and strike demands from EDF agents and their unions.

Power Plant Capacity and Work Schedule Organization Constraints

To ensure nuclear plant competitiveness, nuclear units have to be able to operate to the maximum, particularly in winter when there demand for electricity is highest. This means that it is necessary to concentrate maintenance operations within the shortest period possible and in any case outside the winter months of December-March. This demand of course runs counter to spreading maintenance activity throughout the year. As in any "seasonal" occupation, specialized maintenance personnel are called upon to endure months of forced inactivity and have no freedom to choose their vacation time. These constraints in turn stand in contradiction with EDF personnel status guarantees, if only with regard to organizing EDF personnel employee vacations between April and October.

In both cases, EDF management has managed to shake off these contradictions by subcontracting out maintenance operations. Because subcontracting allows for employment flexibility of a kind prohibited by EDF personnel status, it means that radiation dose can be spread out among persons who do not operate permanently on the sites (while ensuring that these persons do not exceed regulation radiation exposure limits), and strict limits may be placed on the unit shutdown period without disturbing EDF's own personnel.

The 1993 Mandil Report on reforming organization of the electric and gas industry ratified this development by recommending that the state monopoly on energy production be abolished while maintaining the state in its service role of transporting and distributing energy. The report also called for internationalization and exportation. This restructuration made adapting to competition constraints an absolute priority for the new EDF company.[2]

[2]Prime Minister Dominique de Villepin opened EDF capital to public purchase for the first time in October 2005.

The "Partnership" between EDF and its Service Companies

A Progress Charter

Between 1993 and 1997, a "partnership" policy was put in place between EDF and the companies performing services for it. This development culminated in January 1997 with adoption of a "progress charter" signed by the principal and the professional organizations that represent the service companies operating on France's nuclear power facilities.[3] The charter's stated aim is

> working toward quality and perfomance, the essential sources of progress in operating power plants under optimal safety conditions, guaranteeing safety at work and radiation protection for operators, and ensuring that the national industry remains competitive. . . . This partnership shall be framed by a system of qualifying and evaluating service companies, the purpose of which is to ensure that they take up and share EDF objectives, as well as by a contract system defining EDF's and the service companies' mutual commitments.

The charter defines seven general principles likely to contribute to "the quality and competitiveness of the kilowatt/hour":

- transparency of job propositions to service companies;
- developing operators' sense of professionalism;
- improving work plan visibility;
- reducing individual and collective radiation dose;
- improving risk prevention;
- improving working and living conditions;
- follow-up.

The most tangible measure in the partnership involves the signing of pluriannual contracts. The other charter points primarily lay down the demand for high-quality organizing of unit shutdowns without giving any specifications about the means to be used to satisfy that demand. The charter was drafted and signed without consulting representatives of either EDF or the subcontracting companies.

In reality, whereas the partnership policy initiated by EDF was first officially implemented in 1993, pluriannual contracts have only been signed for a small

[3]The charter was signed by the following representative organizations: Fédération des Industries Mécaniques, Groupement Intersyndical de l'Industrie Nucléaire, Conseil National de la Maintenance Industrielle, Groupement des Industries de Matériel d'Equipement Electrique et de l'Electronique Associées, Syndicat des Entrepreneurs de Réseaux et de Construction Electriques, Comité Professionnel des Prestataires de Services en Matière d'Assainissement Radioactif.

proportion of subcontracted work; namely, operations carried out by companies that have direct job deals with EDF. As we know, there is also the practice of sub-subcontracting, where subcontracted companies subcontract part of their work to their own subcontractors. The provisions of law no. 75-1334 of December 31, 1975[4] on subcontracting raise a serious question, and it is worthwhile quoting in full Article 3 of this law, which addresses the matter of sub-subcontracting:

> The company (subcontracted by the principal) that intends to execute a contract or an agreement by means of one or several subcontractors must have each subcontractor accepted by the company that controls and oversees the work, and certify the payment conditions for each subcontract contract at the time the agreement is concluded and throughout the entire length of the contract or agreement. The main company is required to communicate its subcontractor contracts to the controlling company when requested to do so by the latter.

The work agreements reached between companies and sub-subcontractors in nuclear industry maintenance work are so opaque that we cannot say how executive managers of France's nuclear power program apply this article on subcontracting. The decree of August 10, 1984, on *"the quality of design, construction, and running of basic nuclear installations"* (see the *Journal Officiel* of September 22, 1984) permits not only the use of service companies in the nuclear industry but also the subcontracting of certain jobs by these service companies, provided the companies establish the required procedures for monitoring and checking quality. The decree does not specify what those procedures are or clear up the inconsistency between them and Article 3 of the law on subcontracting. Meanwhile the partnership charter of 1997 says nothing about EDF's "partner" companies' subcontracting out its jobs. It only mentions individual operator and service company certification, including that of temporary agencies if service companies decide to use temporary workers. This point is discussed further on.

The objective of the partnership charter was to establish pluriannual contracts for six to eight million hours of work by the end of 1998. A 1997 report by Birraux, member of parliament working at the Office Parlementaire d'Evaluation des Choix Scientifiques and Technologiques[5] indicates that for France's nuclear facilities in 1996, out of a total of 14 million hours, 4.8 million hours worked were covered by contracts (Birraux, 1997). The report does not specify the type of jobs covered by these contracts. However, in reference to the division of labor that we have identified by means of DATR worker accounts, we can hypothesize that the partnership concerned primarily highly specialized work critical to safety, jobs made less dangerous in terms of radiation exposure risk by

[4]This law was later modified by the law 81-1 of January 2, 1981.
[5]The Parliamentary Office for the Assessment of Scientific and Technological Choices.

preliminary opening, preparing and decontamination work—i.e., work done by companies without any claim to pluriannual contracts. This hypothesis seems confirmed by the following account from Birraux concerning the plant constructor Framatome.

Despite a pluriannual contract with EDF, Framatome and its subsidiaries experienced an *"overly heavy work load"* in 1996 that required it to implement "adaptation strategies" using the following three modes, observed to be practiced at Jeumont Industrie (a subsidiary of the Framatome group):

- use of "sedentary" personnel from Jeumont Industrie workshops as support for "itinerant" workers;
- use of seasonal nuclear maintenance workers from among personnel laid off during the labor restructuration of 1993 that had followed Framatome's purchase of Jeumont Industrie. The new management at Jeumont offered contracts for between 1000 and 1300 hours a year distributed over a 12-month period to some laid-off personnel; it thus profited from the nuclear industry skills that had been acquired by these former employees without having to pay full-year unemployment insurance for them;
- Jeumont subcontracted work to other companies.

Furthermore, the company seems to have been organized in terms of what management *"modestly"* called *"fluidity measures"* (as Birraux points out), which meant calling in temporary workers (either workers on limited-time contracts or temps). It should also be noted that two Jeumont Industrie employees were prohibited from entering the controlled zone of the Saint-Laurent-des-Eaux nuclear plant in February 1997 because they were not wearing dosimeters (*Le Monde*, Feb. 5, 1997).

As this example shows, pluriannual contracts have not necessarily involved stabilizing employment or eschewing the use of temporary workers. While EDF indicated to Birraux that use of temporary agencies should remain *"exceptional,"* it also declared that using them was *"sometimes unavoidable."* But no real specifications are given about the type of operations performed by temporary workers on this or that site. And given what Birraux discovered in visiting a Manpower temporary agency near the Saint-Alban plant, it is questionable whether use of such workers is actually exceptional. This multinational temp agency chose to invest in CEFRI certification[6] (to be discussed further on) and the training of 700 DATR temp workers in accordance with EDF norms. The agency would not have done this if it did not expect to recoup the investment. Would a company of that size have taken such a risk for an uncertain market and *"exceptional"* hiring, at a time when demand for temporary workers was

[6]Certification allowing companies to train and do follow-up on personnel whose work exposes them to ionizing radiation.

continually on the rise in all sectors and the company itself had 330,000 *"collaborators,"* 70,000 working under contract at any given moment?

In meetings between EDF and nuclear servitude employers (the current term is now "radiation decontamination") regarding policy for "back-up personnel in response to the seasonal nature of activities," the strategy of using temp workers was confirmed as the only one that could meet cost and flexibility imperatives. Preference was given by subcontractors to temp worker leasing companies for the following reasons:

1. 2/3 of work accident costs fall to the temporary agency, whereas for workers on a limited-time contract the entire cost falls to the subcontractor;
2. if the temp worker goes on sick leave, he is replaced;
3. if work time is reduced or dose limits exceeded, the temporary agency contract is "more flexible" than a limited-time contract.

This strategy and the stated reasons for it indicate not occasional but rather organized if not systematic use to temp workers; indeed, the whole system for managing work and risks is organized around using temporary workers in various ways. This stands in contradiction with Article L 124-2 of the *Code du Travail* stipulating that "temporary work contracts may not have for aim or effect to fill a job position linked to user company's normal, permanent activity."

On May 17, 1999, the Conseil des Prud'hommes (industrial tribunal) of Tours ruled that the work contract of three employees under temporary contract with Polinorsud company had to be redrafted, and it awarded 50,000 francs (approx. 7600 €) in damages to each employee for illegal lay-offs.[7] Cited as grounds for the decision was the company's strategy of "institutionalizing limited-time, and therefore unstable, employment" for tasks that were part of its usual activity. The Conseil des Prud'hommes underlined that "by acting this way, the limited company Polinorsud developed a labor mass on limited-time contract that it intended to use to handle the (usual) 'increase in its activity' across all sites, without ever being called upon to specify what this increase amounted to, the precise task to be performed by these employees, and the moment the increase came to an end" (Article L-122-3-1 of the *Code du Travail*). The company in question worked in nuclear servitude, performing radioactivity decontamination before maintenance operations, among other things. The work contracts concerned "performance of a precise, non-lasting task, nuclear servitude, due to a temporary increase in our activity." The grounds for the judgment specify that the "task" in question were always "cleaning positions"; that is, "dose-costly" decontamination activity in which regulation radiation exposure limits are quickly reached. We have here an implicit legal condemnation of managing hiring and job assignments in terms of radiation dose.

[7] Conseil des Prud'hommes of Tours, hearing of May 17, 1999, report no. 219/99.

Service Company Employers' Understanding of the Partnership

Outside company employers are critical of the forms that the partnership with EDF has taken. The themes mentioned in interviews with managers of these companies echo the results of a study commissioned by EDF[8] and responses given to Birraux in 1994 and 1997.

According to Birraux (1994):

> This policy is due to a reversal of nuclear plant operation principles: from a system of controlled spending that required the service company to use limited resources, we have moved to package-deal orders that require the company to produce results. In areas where quality is directly at stake, this requires skilled operators, tested methods and instruments, and highly technical oversight and verification). (p. 236)

This is what justifies the "partnership," defined in principle as putting a number of customer-supplier mutual commitments under contract.

In the view of service company managers and overseers, relations with EDF are characterized by 1) discrepancy between discourse and practice, and 2) a contradiction between the goals imposed and the resources for reaching them.

The discrepancy between discourse and practice is manifest in EDF purchasing policy. Despite its discourse on safety, it seems that financial and business relations criteria have priority for EDF, to the detriment of technical quality. As one subcontractor manager put it, awarding jobs is really "the prince's prerogative," in this case that of the EDF services that choose which candidate to award the job to, within the bidding framework but usually independently of the criteria inscribed in the Charter.

Moreover, while the partnership creates demands and obligations for service companies seeking to attain relative stability in the form of an several-year agreement with EDF, EDF itself is not tied down by obligations toward its "partners," and it can challenge a pluriannual contract, citing organization decisions made with regard to maintenance. In fact, starting in the second year of such a contract, EDF can unilaterally and without any compensation reduce by half the annual work-hour package that a company was granted.

Lastly, the discrepancy may be felt in the conditions that operations are performed in, particularly in relation to safety and radiation protection goals.

Birraux (1994) noted the same perception of the partnership by service company employers. He notes the absence of real obligations for the principal and lack of any negotiation latitude for service companies on deadlines or the package-deal principle for operations, while underscoring the negative consequences of making unit shutdowns seasonal:

[8]SORGEM-EDF/PT, 1995, *Etude des relations avec les prestataires,* Paris (written report of an oral presentation).

> low availability of company equipment and men. This implies tensions around production apparatus and personnel. Seasonalizing unit shutdowns seems to be the main cause of using temporary, unstable employment.
>
> high training costs for operations in the controlled zone, costs not recouped by activities during off-peak periods. This problem is critical for highly specialized personnel who may be approaching the dose limit.
>
> rapid loss of "nuclear-style" precautions when workers have not been operating in the controlled zone for some time. Moreover, it seems difficult under these conditions to get personnel to follow an ALARA approach, which is by definition long-term.[9]

The contradiction between imposed goals and the means for reaching them is particularly sharp with regard to management operation time and the problem of radiation exposure. Seasonalizing and shortening unit shutdowns has made it extremely difficult to remain within legal work hours.

> The impression we have is that we're always being asked to do more: constantly on the red line with regard to labor legislation and at the same time, there's a two-sided discourse on quality and safety that is incompatible with hard-line economic profitability (manager of a subcontracting company)
>
> The change has been for the worse—that much is certain. We're being told to get the work done in shorter and shorter periods, because unit shutdown times are decreasing, and to keep technically skilled people while obeying labor legislation. . . . That multiplies demand for temp workers who have received little or poor training because no company can work twelve months for work that has to be done in six, or else you have them work 150 days in six months and you've broken the law. (manager of a subcontracting company)

EDF's refusal to pay more for subcontractors that would use techniques aimed at reducing dose goes against the ALARA approach, i.e., decreasing radiation exposure to the lowest possible level:

> ALARA went through on paper, but I've seldom seen EDF pay more for a technique that would allow for gaining on dose. What they say is "Yes, that's an interesting technique, and I want it for the price we paid before." (manager of a subcontracting company)
>
> We're going to hit a limit, a point where gaining on dose will become contradictory with reducing costs. (manager of a subcontracting company)

[9] Approach for optimizing dosimetry: the point is to receive only "worthwhile" radiation at the "lowest possible level." Radiation protection standards and practices are analyzed in chapter 5.

In his 1997 report, an assessment of the reform begun in 1990, Birraux underlines the persistence of "contradictory messages" in regard to EDF's purchasing policy (the contradiction between "cheapest" and "best quality" offer), the intense time pressure on EDF agents ("How can a project manager work 48 hours a day?"), the upsetting of power relations such that service companies are now entirely responsible for site operations, and the continuing policy of accepting the "cheapest offer."

A combined economics and labor law study of the concrete modes by which subcontracting bids are awarded for different levels of maintenance during unit shutdowns would provide precise information and real transparence with regard to current practices for organizing maintenance work in nuclear power plants. To my knowledge, no such study has been undertaken, and there is no regulatory arrangement, even at the level of state public oversight, that provides access to this information.

This means that even though the way subcontracting relations are organized has changed in the last ten years, the strategies adopted have not changed the principle of domination that governs the partnership between EDF and service companies. We can even hypothesize that these strategies are strengthening EDF's power over the labor and technical organization of outsourced work, without loosening the grip of the economic constraint. This analysis is similar to Hancke's in his study of changes in subcontracting relations in the energy and automobile sectors in France compared to other European Union countries (1998). Questioning analyses of the partnership that represent it as a relation of "trust" between principal and subcontractors, he shows that this hypothesis cannot resist study of what remains a power relation, fundamentally asymmetrical despite changes in the way it is enacted over time.

Nuclear Service Company Employees' Charter of Demands

Echoing the nuclear industry management charter, the CGT unions at EDF and in the subcontracting companies working at the Chinon power plant developed a charter of demands in 1998 for service company employees working in the nuclear industry. This charter shows the gap between the stated principles of the EDF/service company partnership and the lived reality of these employees. Some of the demands very directly concern the modes of "partnership" between EDF and service companies. It calls for:

- Another way of planning EDF maintenance work (not limited to unit shutdowns or the nuclear industry). Planning must spread the workload over the entire year. No operations should be programmed at night or on the weekend.
- No more "package-deal" contracts.
- A single status for all operators, regardless of company and occupational branch. This status needs to include qualifying training programs, allow for

free choice of vacation time, guarantee that all time spent on site is remunerated, take into account and reimburse real travel expenses.
- Prohibiting use of temporary, unstable employment for all operations exposed to radiation.
- Immediate application of the dosimetry limit of 20 mSv per year, with a maximum of 2 mSv per month. An employee who reaches this limit needs to be maintained entirely in his job and continue to receive full pay. The financial cost of this measure should be borne by EDF, which is responsible for the risk.

These demands, based on rights recognized by the *Code du Travail* concerning working hours, training, vacation time, and health protection in the workplace bring to light the contradictions between the nuclear industry partnership as applied in practical organization of maintenance work and the full exercise of these rights.

Toward Conditional Maintenance

In 1997 a new concept "conditional maintenance" began to appear in memoranda from EDF management to EDF personnel. Conditional maintenance involves reducing the number of systematic preventive maintenance operations. To determine how to do this a study was conducted in 1997 (Berneau, 1997) on the initiative of France's nuclear program managers.

"The nuclear program authorities view conditional control as one means of reducing maintenance costs. The goal of reducing costs is itself inescapable if the competitiveness of nuclear power is to be guaranteed over the long term," explains an EDF manager (Pesteil, 1997). While technical "optimization" criteria have their place in strategies of conditional maintenance, the main goal appears to be reducing maintenance costs by 8 to 20%. The 1997 activity report, by nuclear program authorities, indicates that "comparing maintenance costs among production units" will play an important role in "generalizing good practices." Clearly there is an obligation to produce results, in this case by making the different nuclear production units compete against one another, and the means ("good practices") have to be adjusted to the competitiveness criteria.

What the documents did not say about the new maintenance policy was the impact it would have on hiring, working, and safety conditions for "outside" workers, on the one hand, and on nuclear installation safety on the other.

Competition has thus imposed itself as the predominant legitimacy criterion for subcontracting maintenance work in the nuclear industry. This development has occurred without major conflict between management and EDF unions. The unions did not in fact oppose extending and generalizing the subcontracting of maintenance work.

Another source of legitimacy, concerning nuclear plant safety, has worked to favor this development; namely, the standardizing of training and quality control procedures in connection with what is called "nuclear safety culture."

THE QUALITY-SAFETY POLICY AND RISK-PREVENTION

Forefronting cost rationalization can only be a credible policy if it does not seem to threaten nuclear safety, at least not officially. Nuclear safety has been and continues to be the main argument for continued development of France's nuclear energy program. It is one of the foundations of consensus between management and the unions at EDF for the choice of nuclear energy as France's nearly unique energy source. The legitimacy of this choice rests on the guarantee that safety remains at the heart of work organization in the nuclear power production system. Safety is understood to depend on quality, a label acquired by any company that complies with quality assurance standards. What are these?

In the industrial system as a whole, quality and its normative aspect, involving compliance with standards, are now at the center of work organization, both on the technical side and in matters of managing persons engaged in a productive activity. Sociologists have studied the meaning and practices of approaches to quality in business, while legal specialists have looked into the relations between such standards, which are a matter of technical rationality, and legal rules, which of course pertain to law.

To understand the meaning of quality-safety policies for maintenance operations in basic nuclear power plants operated by EDF, it is worthwhile looking in some detail at sociological and legal analyses of "quality" and standardization in the French and European Union industrial contexts.

QUALITY ASSURANCE AND ISO STANDARDS: RATIONALIZING PRODUCTION OF GOODS AND SERVICES

Mispelblom (1995), Dassa and Maillard (1996), and Rot (1998) have done sociological studies of the history of "quality" and related discourses and social practices in the goods and services production system.

Challenging the concept of quality as falsely universal, Mispelblom has shown how it is in fact a social construction that evolved over the course of industrial history. Quality emerged as the guiding philosphy of Taylorism in the form of a statistical instrument for assessing production performances; it then became an ideological system used by employers and principals, the first vis-à-vis their employees and the second vis-à-vis subcontracting companies and *their* employees, to impose rules, execution procedures, and means of checking that these rules and procedures are followed. Above all the author shows how, through the quality assurance ideology, value has been displaced from "a job well done" to ensuring conformity with ISO 9000 standards, quality standards developed by expert commissions and codified according to an internationally recognized and accepted procedure in the framework of the International Standards Organization. This definition of quality—as compliance with and conformity to procedures—has become the cornerstone of business quality assurance procedures and different types of business certification. Mispelblom presents the ISO standards as an instrument for institutionalizing the demands of European competition, thus creating what he calls the "certified club"; that is, companies that have been judged by expert certification organizations to comply with standards adopted at the international level. In a wide-ranging bibliographical overview of studies of certification procedures, Mispelblom brings out how these procedures have tended to confirm the hierarchy of production tasks (and businesses) and distinguishes three types of standards related to procedures: design, realization and checking.] "The general aims of this quality pyramid are summed up by the increasingly widespread slogan 'Say what you do, do what you say, and prove it'" (Lamprecht, 1994). "The 'you' refers here to both the company as a whole and the individual personnel member, who has to do what he is told to do in accordance with detailed instructions that also indicate how they are to be followed. ISO standards not only require goal definition; they also insist on the means used to reach goals, through written procedures" (Mispelblom, p. 63).

The fact is that "defining quality as conformity to existing procedures implies that those who design the procedures are assumed more or less infallible" (Mispelblom, p. 95). This also reaffirms the forms of legitimacy and power already operative in the production system, legitimacy being on the side of those who develop standards and procedures, while the real work of those who apply them has to conform to those standards and procedures regardless of the fortuitous occurrences encountered in the activity itself. This is the "fantasy of a programable business" (p. 166) which, in the view of management, itself understood as the science of "human resource" management, requires using the quality assurance method as an instrument for rationalizing social organization of work. Mispelblom adds that at the scale of the European Union, "confidence in the European label is tending to replace the trust that was established on the basis of regional or family traditions that involved ancestral know-how and traditional relations in which a spoken promise was sometimes considered more solid and inviolable than a written contract" (p. 167).

At issue in quality assurance and the set of procedures known as "total quality" are not only standards and procedures for design, realization, and work inspection or checking, but also standards for personnel behavior. The latter involve a sort of moral code for the "customer-supplier" relation, particularly in the organization of subcontracted work. The point is "general mobilization" of all in the service of an "excellence" project that will guarantee that the real quality of the work corresponds to the certified quality—in order to satisfy the customer. Mispelblom shows how this new work morality necessarily makes divergence from labor law rules necessary and legitimate (rules covering such matters as work hours or schedules and work conditions or intensity) as soon as the customer's order, in conformity with imposed quality standards and procedures, requires such divergence. The question of the conditions and constraints to be borne by workers (generally designated "operators" and even "collaborators" or "partners") during the work activity itself is not raised. It goes without saying that conformity to standards not only presupposes that operators adapt themselves to these demands, but also makes them responsible for meeting them, as Rot shows in her study of the automobile industry (1998). The author brings to light the fundamental contradiction in quality assurance between transparency and responsibility, which in assurance logic can be expressed as "imputability." Caught in this contradiction, operators "do not necessarily comply with the traceability imperative and are frequently led to find compromises between the demand for justice (they will not 'note down' and therefore 'denounce' an operator who is in fact seldom fallible) and a demand used by the company as a governance instrument; namely, following the rules: 'everything must be recorded.' ... This means that in the construction of immediate local arrangements, informal regulations play a non-negligeable role" (Rot, 1998, p. 13).

In a study commissioned by the food industry, Dassa and Maillard studied businesses' strategies and operators' practices in relation to quality policy in businesses producing and packaging fresh foods (Dassa & Maillard, 1996). Their analysis is based on observation of both the rules, standards, and practices involved in the companies' "hygiene-cleanliness-quality" programs, and the actual conditions and organization of work. The authors bring to light the assumptions underlying the food industry's request for the study. They explain that they were expected to "analyze the poor results of training programs that purported to lastingly change personnel behavior in the areas of hygiene, cleanliness, and quality." They explore the contradictions in work organization between the rules promoted by the quality program and real working conditions.

The assumptions that the authors bring to light reflect an instrumental idea of human labor as productive factor, and idea that "management science" rationalizes the use of this factor to improve business productivity. The psycho-sociological assumption concerns employees' "natural" adherence to such a program: it goes without saying that employees should make the program rules their own and spontaneously commit themselves to applying them. According to the "social assumption (prejudice) that stigmatizes task-performing personnel," the workers themselves are responsible from the outset for the project's failure. They are assumed to have closed their minds to any effort to change; managers do not even ask about the characteristics of the social organization of labor and social relations within the company. Their cognitive assumption interprets this failure in terms of cultural ignorance: employees are viewed as foreign to the culture of cleanliness. This assumption is based on the hypothesis that knowledge of a risk automatically brings with it a change in behavior with regard to that risk. Here again, the managerial understanding of human resources management is blind to the history and forms of socialization in which the genesis of individual and collective behavior with regard to risk—and changes in that behavior—are rooted. That understanding is equally obtuse with regard to cultural processes and social differentiation of representations and practices around the question of "clean" and "dirty." According to a didactic assumption closely linked to the cognitive one, changes in behavior only come about through specific and adequate operator training—as if the didactic message carried within it the power to make people change. The authors underline that in fact "the power [to change] comes from the social and institutional framework that imposes or presupposes such change and the social relation instated in this connection."

Together with the assumptions on the part of management, Dassa and Maillard's results for the three sites they studied bring to light "the weight and primacy of organizational factors as sources and justifications of behavioral practices, whether these comply with prescriptions or not." Several types of organizational factors are relevant here, including job management policy, wages and working conditions in the company, the status of "outside" workers (subcontracted employees, temp workers) and the forms of social control over individuals that hygiene-cleanliness-quality programs tend to establish. As the authors show, given that this project carries with it the "prescriptive utopia" contained in all certification approaches of the ISO 9000 sort, the main objective tends to be following rules and procedures, regardless of any contradictions that may appear: contradictions between economic logic and the social logic of applying the rules; between that social logic and the specific characteristics of the material environment and productive tasks themselves; and between prescribed procedures and know-how acquired through experience on the job. This know-how is no longer what gives the operator personal assurance (in the sense of mastery) when performing the work activity. On the contrary, if ever the know-how has led the worker to develop practices that do not fit with the procedures, he or she has to conceal them in written accounts of the compliance of his or her work to the prescription. Written proof of practices' conformity to prescriptions becomes more important than the resistance that real work offers to being grasped by what Dassa and Maillard call "procedural reality."

Standardization and the Law

The preceding analyses emphasize the philosophy behind standardization: the goal of standardization is not the quality of products or services but conformity of equipment, procedures, and behavior to standards established by private expert commissions. In this sense it may be linked to insurance philosophy, which dissociates the insured risk (and modes of compensation when that risk is realized) from the causes that give rise to the risk (accidents, water damage, natural disaster).[10] In ISO 9000 certification and the different types of standardization, emphasis is not only on the intrinsic quality of the product but on conformity to standards. Conformity is attested by the person whose responsibility it is to write that the work has been done and that the product thus produced conforms to the standards. If a discrepancy exists between the product and the declared quality standards, the responsibility for this is to be imputed to the operator, caught "redhanded" not following the rules inscribed in these standards. Real work modes (which are by their very essence non-programmable) cannot be opposed to standards assumed imperative for obtaining quality certification. Standards, which have developed out of certification procedures, thus tend to become rules pertaining not to technical evaluation of points of non-conformity but rather to legal logic (in the sense of a legal obligation) and transgression of that logic. What, then, are the relations between these standards and labor law? The development of standardization at the European Union level has led jurists to look into these relations.

Standardization activity has intensified in Europe over the last twenty years, namely in the area of occupational safety with regard to machines and equipment. In the debate on regulation harmonization at the European Union scale, jurists have taken up the question of the relations between legal rules and technical standards in this regard. This debate allows for examining both legal rules and technical standards that have been developed as well as the value systems operative—implicitly in most cases—in that development.

Without entering into the complex dynamic of the EEC's process for developing legal rules and technical standards, we can understand the expansion of standardization activity as an instrumentalizing of law in the service of the free movement of commodities within the social and labor space of the EU. Debate on harmonizing social and labor legislation has led to a twofold approach to adopting rules and standards that are to be shared by the different EEC states. When this approach, called the "new approach," is applied to safety, it combines directives (legal rules of a public character) with standards. The definition of standards given by Vacarie and Supiot (1993) is "technical specifications approved by a recognized norming organization and designed to be applied repeatedly and continuously, in such a way that observation is not required." Directives, meanwhile, "state essential requirements and leave to private standardization organizations the task of developing the technical specifications that will meet those legal requirements." Technical specifications become standards which are then incorporated into law.

Vacarie and Supiot present a critical analysis of this approach and the modes of expertise specific to standardization. In their understanding, there are two

[10] Cf. François Ewald's 1986 analysis of work accident law and the creation of a social protection system in the framework of the welfare state.

different systems of legitimacy, one that founds legal rules, the other technical standards. The technique of "referring to standards" is consistent with a divison of labor between the public and private spheres that is part of a general movement in social and legal labor organization promoting "business self-regulation": "technical standards are developed by private bodies created by business, bodies whose expert assessment activity is not subjected to any form of public oversight and in which, even in matters of occupational health and safety, there is no provision for worker representation."

Still more fundamentally, what the authors are interested in is the difference between "legal rules" legitimacy and "technical standards" legitimacy. The legitimacy of technical standards comes from a "knowledge of facts" (knowledge essentially modifiable and created by means of theoretical systems of interpretation that are forever temporary, and methods for approaching these facts) that refers to scientific and technical rationality.[11] "Legal rules" on the other hand, "draw their legitimacy not from knowledge about facts but from values, to which they intend to subject facts." The EEC's "new approach" to harmonizing labor law breaks with this fundamental difference and the labor regulations deriving from it. Law is instrumentalized in that "the legal rule is perceived as merely the shape given to economic and labor 'regulations' that draw their legitimacy from scientific and technical knowledge." This knowledge is in fact not only run through with uncertainties but also and above all dominated by conflicts of interest. The authors cite as an example the European directive on exposure to ionizing radiation which specifies that exposure "must be reduced to the lowest reasonably practicable level."[12] As the authors point out, the "reasonably practicable" level will not be the same, or assessed in the same way if it is a matter of cost-benefit relations for a business or the focus of strict application of a precaution principle for the health of DATR workers. Experts in charge of fixing the exposure level do so with what is generally implicit reference to their social and scientific experience, the system of values they adhere to, and the interests they are representing (are they representing the interests of workers' "health" or the interests of the "health" of the nuclear industry?). Standards thus refer not to principles (such as the right to health protection in the workplace) but to a kind of arbitration between such rights and/or interests that are in contradiction with them. The absence of workers' representatives in the process of drawing up standards greatly facilitates employer consensus, a consensus untroubled by the expression of radically contradictory interests. The

[11] "[A]ll science, which is objective in terms of its method and object, is subjective with regard to tomorrow since, short of assuming it to be completed, many of today's truths will become yesterday's mistakes" (Canguilhem, 1966; Eng, 1989, p. 212).

[12] Article 6 of Directive no. 80-836.

standards are then imposed in the name of science, including when they concern performance of work procedures.

Vacarie and Suppiot plead for restrained use of technical standardization:

> As long as the human person (in the legal sense) is at stake, we are dealing with values, an area that the law must not abandon to the power of scientists or technicians, either directly (through reference to standards) or indirectly (through use of 'ethics' committees). . . .
>
> Moreover, reference to standards is only acceptable if the principle of open debate is applied in developing and implementing those standards. If this is not the case, there is no guarantee against economic abuses and technical insufficiencies, in turn likely to contaminate those standards. This points to two imperatives that are either not taken into account or under-estimated in the "new approach." The first is balanced representation of interests in the development of standards. . . . The second is submitting technical standards to the eventus judicii; in other words, opening legal avenues through which their validity can be tested. (p. 26)

This plea went unheeded, swept away by the choice of a standardization policy whose legitimacy was strengthened by the subordination of European Union labor law to the demands of competition. In an analysis of developments in labor law and the weakening of workers' rights in European Union states, Vogel has brought to light a displacement of the sources of law legitimacy. He cites a report by the Molitor Group to the European Commission on "legislative and administrative simplification," a report which states that "the main source of legitimation for a legal standard is its economic effectiveness, and every legal standard must be analyzed in cost-benefit terms that demonstrate its positive impact on economic development" (Vogel, 1997). Standardization implicitly carries the same legitimation criteria. These criteria, which were not discussed in open debate in the public space, have imposed themselves with particular strength because they are draped in the appearance of experts' scientific and technical rigor and objectivity.

We now consider how the nuclear industry's current way of managing overall safety, quality and worker safety has acquired undisputed social legitimacy.

Safety, Quality Assurance, and "Safety Culture"

Given the dangers inherent in the production of nuclear energy, safety in this industry rests on infallible control of accident risk—using "a deterministic approach," say the texts. Who is responsible for this control? What guarantees are given of the infallibility of such absolute control of nuclear accident risk? How is the "quality" of maintenance work, 80% of which is done by "outside" workers, guaranteed? Above and beyond the technical issues involved in safety, one fundamental question comes to the fore: Who assumes responsibility for "infallibility" and how?

Nuclear Safety

> The size of nuclear operations in France and the nature of the risks they involve have led the public authorities to develop a specific system for organizing and regulating nuclear safety. The fundamental option that this system is based on is owner's technological responsibility. The public authorities shall ensure that this responsibility is fully assumed in accordance with regulation prescriptions. (Direction de la Sûreté Nucléaire or DSIN [French nuclear safety authority], 1995)

It is therefore the owner company that has to assume technological responsibility for establishing and implementing safety measures designed to meet the general objectives defined by the public authorities. Safety is founded on the principle of "deep defense." This consists in "layered provisions (lines of defense) for countering technical or human failure or error."

To guarantee safety, EDF must guarantee the technological reliability of installations, but also "be certain of" the human conduct and reactions of operators intervening in the areas of both production and maintenance. The nuclear industry offers an example of a particularly detailed approach to this pair of dimensions in industrial management.

The French ministry of industry decree of August 10, 1984 (published in the *Journal Officiel* of Sept. 22, 1984) "on the quality of design, construction and running of basic nuclear installations defines the general framework of owner-operator responsibility concerning quality with regard to ensuring safety of basic nuclear installations." Formal, highly detailed procedures for overseeing and verifying the conformity of activities to the required quality standards were instituted, specifying intervention procedures, prescriptions, and instructions in-depth. Anomaly and incident analysis was codified and made subject to examination by the nuclear safety authorities. The decree permitted recourse not only to subcontracting, but also to sub-subcontracting as long as supplementary oversight procedures required for quality checking are put in place. Operators' qualifications must also be verified according to particular procedures.

These provisions institutionalized written "traceability" of operations and confirmed the dissociation between actual task performance and oversight. The written "traces" of the operation thus became the means of attesting that it was performed and designated the person(s) responsible for it. Quality assurance involves verifying that "traceability" conforms to the quality standards defined in the framework of the national and international standardization procedures specific to the nuclear industry.

However, the loss of direct control over task performance, the number and diversity of operators (EDF agents and subcontractors), the extreme complexity of the division of labor in production and maintenance also required that a new kind of influence be exercised on the operating modes of those in charge of maintenance. The obligation to produce results, obligation sanctioned by the document by which an operator commits himself, does not provide an absolute guarantee of the quality of human intervention. Nuclear safety thus involves another dimension of standardization; namely, the attitudes and behavior of operators working on nuclear installations.

The Safety Culture

The shock provoked by the Chernobyl accident led the International Atomic Energy Agency (IAEA) to define behavioral standards for a "safety culture" meant to be a system of reference for all workers operating in nuclear power plants (see International Nuclear Safety Advisory Group [INSAG-4] 1991). Starting with the hypothesis that "One of the most important lessons of abnormal events, ranging from minor incidents to serious accidents, is that they have so often been the result of incorrect human actions", the International Nuclear Safety Advisory Group (INSAG-4) commissioned by the IAEA to develop the nuclear safety culture, specifies as follows:

> The phrase "safety culture" refers to a very general matter, the personal dedication and accountability of all individuals engaged in any activity which has a bearing on the safety of nuclear power plants. These matters are especially important for operating organizations and the staff directly engaged in plant operation. For the latter, at all levels, training emphasizes the significance of their individual tasks from the standpoint of basic understanding and knowledge of the plant and the equipment at their command, with special emphasis on the reasons underlying safety limits and the safety consequences of violations. Open attitudes are required in such staff to ensure that information relevant to plant safety is freely communicated; when errors of practice are committed, their admission is particularly encouraged. By these means, an all pervading safety thinking is achieved, allowing an inherently questioning attitude, the learning from experience, the prevention of complacency, a commitment to excellence, and the fostering of both, personal accountability and corporate self-regulation in safety matters.

The concept of safety culture is refined in 75-INSAG-4, which specifies requirements on three levels:

> *Requirements at policy level*
>
> In any important activity, the manner in which people act is conditioned by requirements set at a high level. The highest level affecting nuclear plant safety is the legislative level, at which the national basis for safety culture is set. Similarly, an organization pursuing activities with a bearing on nuclear plant safety makes its responsibilities well known and understood in a safety policy statement. This statement is provided as a requirement to managers and staff, and to declare the organization's objectives and the public commitment of corporate management to nuclear plant safety.

> *Requirements of Managers*
> The attitudes of individuals are greatly influenced by their working environment. The key to an effective safety culture in individuals is found in the practices moulding the environment and fostering attitudes conducive to safety. It is the responsibility of managers to institute such practices in accordance with their organization's safety policy and objectives.
>
> *Response of Individuals*
> The response of all those who strive for excellence in matters affecting nuclear safety is characterized by a questioning attitude, plus a rigorous and prudent approach, plus communication. The desired results are achieved only if the attitudes of individuals at all levels are responsive to the safety culture framework established by management.

For employees, the emphasis is on the importance of evaluation not only of technical skill but also behavior.

> Operational excellence is achieved in present and future nuclear power plant operations by: augmenting safety culture and defense in depth; improving human performance; maintaining excellent material condition and equipment performance; using self-assessments and peer reviews; exchanging operating experience and other information around the world; and extending the implementation of severe accident management.
>
> . . . Most of the preceding enhancements are aimed at improving human performance since a large fraction of plant events are caused by human error. Errors may be the result of the behaviour of a single individual, the collective behaviour of individuals, or the influences of the work environment, the organization or the management. Excellence in human performance is attained when all the individuals involved exhibit desirable behaviour and when the emphasis is put upon fostering or displaying a questioning attitude; reinforcing such desirable behaviour; encouraging leadership and promoting teamwork.

Individual workloads are thus considered by the IAEA a major, highly influential safety factor.

France's nuclear facilities Authority officially adopted "Safety culture" in the 1990s. A 1994 report by the taskforce on human factors in the department in charge of running France's nuclear facilities presents a synthesis of ideas on moving the "human factor" toward professionalism and a safety culture. The report underscores the move from a mechanistic conception of human intervention to one that makes humans feel responsible for their actions. The first kind of approach presupposes strict compliance with determined, prescribed operating modes, whereas the second acknowledges the fact that human initiative is necessary for quality work and resolution of problems, incidents, and accidents. The first approach is based on cognition, whereas the second pertains to ethics. The report nonetheless forefronts the following paragraph from INSAG-4:

> Accident prevention is the first safety priority of both designers and operators. It is achieved through the use of reliable structures, components, systems and procedures in a plant operated by personnel who are committed to a strong safety culture. . . . However, in no human endeavour can one ever guarantee that the prevention of accidents will be totally successful. Designers of nuclear power plants therefore assume that component, system and human failures are possible, and can lead to abnormal occurrences, ranging from minor disturbances to highly unlikely accident sequences.

According to the EDF report, residual risk is ineluctable, and the report underscores "human failure," with the idea that the "safety culture" tries to limit such failure, particularly by distributing rewards or punishments that are necessary if attitudes in conformity with safety culture are to be imposed:

> A reward is an act of recognizing someone (recognition of a result, i.e., a contribution to the interests of the business, or of demonstrated allegiance) that is either attested publicly or not. Punishment is the opposite of reward, and marks disapproval, either public or not. Punishment should be enough of an attack against the person in question as to make him question himself and change. The idea is to play on his feelings of pride and guilt. (DEPN-FH, 1994, p. 81)

Training and a reward and punishment system are considered especially effective instruments for guaranteeing the development of "safety culture"—understood as a moral attitude that every operator working in a nuclear power plant must adhere to. This tends to transform the labor and wage relation into one of allegiance to the principal's interests, without any legitimate space for critical discussion or negotiation of the organization of work, working conditions and resources for meeting the material demands that would effectively ensure the safety of nuclear installations.

Certifying Companies to Train and Follow-Up on Personnel Working Under Exposure to Radiation

In 1990, the *Groupement Intersyndical des Industries du Nucléaire* (GIIN; interunion nuclear industry group) created the *Comité Français de Certification des Entreprises pour la Formation et le Suivi de Personnel Travaillant sous Rayonnements* (CEFRI), a committee for certifying companies employing personnel directly assigned to work under exposure to radiation and training institutions specialized in risk prevention. This committee is made up of owner-operators/ (EDF, CEA [*Commissariat à l'Energie Atomique*], COGEMA and the French army's weapons department), *the Société Française de Radioprotection, the Office de Protection Contre les Rayonnements Ionisants*, company occupational physicians, and representatives of the *Caisse Nationale d'Assurance Maladie*, France's national health insurance program. Though it has no official legislative existence,[13] it determines the objectives, rules, and constraints of CEFRI certification, certification which the owner-operator EDF has made compulsory for training institutions and temporary agencies that send temp workers to subcontracting companies operating in nuclear plants.

Certification consists in ensuring that training procedures and DATR worker follow-up conform to CEFRI standards. It is granted after an audit performed by experts appointed by the committee. The company requesting certification pays for the audit and certification procedures. Certification must be renewed every four years.

Quality assurance is thus done from the outset according to formal procedures and not the real experience of work and working conditions. Companies and institutions are certified before they operate in a nuclear plant. The presence of representatives of public institutions on the committee (OPRI and CNAM) creates confusion between the company manager's safety obligation as specified by the Code du Travail (and how it is implemented in real work) and the procedural standards for work prescriptions as developed by the nuclear industry. The participation of these bodies works as a kind of "public" approval of an approach that places the issues of exposure, prevention, and health surveillance as they concern nuclear workers in a private space, outside the union or parliamentary bodies in which labor laws and legislation are debated and adopted. Certification thus substitutes compliance of companies' stated intentions to standards based exclusively on the nuclear industry's own expertise for application of labor laws. In Vacarie and Supiot's terms, this is "business self-regulation" which, in this case, involves a high number of blind spots as far as radiation protection law is concerned, as we shall see further on.

Two decrees of December 24, 1998 now make it compulsory for subcontracting companies and temporary agencies operating in nuclear controlled zones to obtain certification. The CEFRI was chosen as a regulatory body to deliver such certification. And the nuclear industries have obtained public legitimacy for "self-regulation" from another private body, the COFRAC, mandated by the public authorities to deliver certification to companies.

Training

The need to obtain outside workers' compliance with its quality-safety policy led EDF in 1992 to make systematic training of employees compulsory for service companies, not only in the area of radiation protection (which was already

[13] This body may be compared to the Comité Permanent Amiante (permanent committee on asbestos) created by the asbestos industry and a public relations agency with the purpose of developing a "culture of controlled asbestos use." For decades this informal structure enabled the asbestos industry to resist against public socializing of knowledge about the pathogenic effects of asbestos, particularly the cancer risk, known within the industry since the 1930s.

required for DATR accreditation) but also in relation to the quality-safety policy itself. Moreover, the basic safety training program was extended to include classic worker safety issues (accident and fire risks, among others) in the framework of "risk prevention" training programs. Retraining programs were also instituted; operators were required to renew their "service company quality-safety" qualification (QSP) and risk-prevention (PR) training every three years.

The worker's company pays for these training programs, compulsory for all DATR outside workers. They are run by training institutions or companies and closely overseen by the CEFRI.

- "Risk prevention" training is in two parts: classic nuclear safety, and related prescriptions that concern primarily accident, fire and explosion risks. It is also in this framework that the characteristics of nuclear power plant operations are explained (the different formalities, instructions, procedures), with modes for identifying the risks associated with operating on a nuclear site and identifying responsibilities—owner operator's, subcontractors'—with regard to worker safety. Very specific information is given about an essential document: the "specifications sheet." This document is the link between EDF inspectors and outside workers; it designates the category and circuits that the given operation fits into and what technical safety arrangements have been made by the EDF agent in charge of establishing the specifications sheet. It includes "key points" that the operating team is under absolute obligation to check before signing the document; signing means that the team in question guarantees the document's conformity with the operation specifications.
  - Radiation protection training provides basic notions about radioactivity, the different measures, the difference between radiation and contamination, borderline values (including the TLV recommended by the CIPR 60), and how monitoring machines work (films, dosimeters). During the training, emphasis is on each actor's individual responsibility with regard to the ALARA policy. Radiation protection is defined as "having the appropriate material, suit, and attitude."
- QSP training trains DATR workers in "safety culture"; that is individual responsibility with regard to the quality of operations and the importance of following the monitoring and verification procedures for those operations. In the course of these training programs, information is given on the technical procedures used in nuclear plants, the different reactor and circuit components, operating modes, and acronyms that workers need to know in order to identify positions and operations in the plant. This material is to be memorized by trainees.

Safety is defined in these programs as:

Reliability of materials + work quality and workers' professional aptitude.

The aim of the training is to make trainees capable of

- adapting their behavior and attitude to fit the quality-safety policy;
- identifying any malfunction or difference from the normal and being able to report on it;
- learning from and contributing to the "return on experience."

The quality-safety policy requires risk analysis and the defining of a "quality plan" that integrates the principles of quality assurance, namely

- prior definition of requirements,
- verification that requirements are followed,
- formalizing [in writing] the physical acts performed.

Clearly the point is to attain formal conformity of maintenance operations through a series of documents that provide a written "trace" that quality prescriptions are followed before, during, and after the operation. Should "particular provisions or means" (DMP) be used that were not planned for in the procedures, a detailed account of what was used must be written up. Trainers underscore that DMPs are dangerous and that their use must be kept to a minimum.

"Training sites" for simulation exercises are used, sites that include "traps" which workers have to be able to identify. Training sites seek to develop three features in trainees: a questioning attitude, strict vigilance and communication abilities.

One area of application used as an example with trainees is site cleanliness. Every service company employee is supposed to feel concerned and responsible to EDF if the site is not clean or other anomalies are discovered during the operation, even if this is due to others' behavior. The employee is responsible not for resolving the problem but signaling it to EDF agents. In the framework of the "customer-supplier" relation between EDF and subcontractors, this "responsible" attitude on the part of subcontractor employees is considered extremely important and even constitutes a "selling point" for a company seeking to work for EDF.

Workers are tested after taking the training program, and if they pass the training is validated. Tests generally require trainees to produce memorized material. Validation is marked on a card that allows them entrance to the nuclear plant; this document is required to get in. Employees also have to pass a test to gain entrance to each individual plant; these tests are to make sure that DATR outside workers do not have memory problems. In certain plants the test is taken on a computer.

These "generalist" training programs do not cover any of the areas of technical qualification specific to nuclear industry maintenance jobs. The subcontracting companies therefore also have to ensure that their employees have received the technical training specified and required by the nuclear plant owner-operator (this is called nuclear accreditation). This requirement makes it particularly difficult for these companies to diversify their activity outside the nuclear field, especially since, to maintain their qualification, employees have to operate at least once a year in a power plant.

The dissociation between "company quality and safety" policy (QSP) and technical qualification represents one of the ways that prescribed and real work are cut off from each other in the nuclear industry. Dejours' study of work-related psychopathology (1992) in the Chinon nuclear plant has brought to light a major discrepancy between the theoretical discourse on safety of French nuclear facilities program superiors and the experience of EDF agents themselves in their daily work, which constantly forces them to "cheat" with regard to the prescribed procedures:

> The intermediate operation supervision level is caught between a rock and a hard place: on the one hand there is the declaration of real technical control of the process (a declaration admittedly more ideological than factually demonstrated), with the requirement that the declaration must be confirmed; on the other, the experience of an absolute, irreducible necessity to cheat and make necessary empirical readjustments in the course of work which is supposed to be strictly organized. (Dejours, 1992, p. 38)

In the relation between technical standards and behavioral standards as those standards have been defined by engineers and managers, operators tend to be blamed for any non-compliance with norms as if they had cheated. Mispelblom (1995) emphasizes this aspect: "With ISO standards, the way work is really handled becomes non-legal" (p. 174). The fact is that from the operators' viewpoint, such "cheating" is more a matter of "informal regulations" such as those noted by Rot (1998), regulations that are indispensable to getting the prescribed job done.

The safety discourse is based on the postulate that outside operators lack the proper serious attitude and spirit of responsibility, a postulate which then is used as the sole explanation for incidents: "human failure" is used to cover up the failure of technology itself and the responsibility of organizational factors. According to Dejours, this criticism of workers' sense of responsibility (directed at both EDF agents and outside company operators) is "a way of concealing the fact that control of the process is imperfect, a fundamental challenge that is hard to bear psychologically."

The subcontracting of maintenance operations play a fundamental role in how this fundamental challenge is concealed from view, because it distances the real work performed and the return on experience for work done directly on nuclear installations. Standardization of work rules is based not on operators' concrete experience but on what that experience *should be* according to the quality-safety assurance idea. We can hypothesize that written accounts of operation results produced by outside workers constantly put them in the position of the accused, forced to justify any and all bending of the rules, rule-bending indispensable if the tasks are to be accomplished, but continually repressed as violation of the rules. The training received in the QSP and "risk prevention" programs makes outside company employees responsible for imperfect control of the nuclear production system and maintenance, and for managing the associated risks.

What System of Values is Operative in Ensuring Nuclear Safety?

Safety culture in the nuclear industry provides a particularly clear illustration of the two-part displacement that standardization brings about in production organization:

First, there is the focal point of quality monitoring and inspection. Quality inspection is not done in reference to the work activity itself or in order to comprehensively analyze the discrepancies between prescribed and real work in such a way as to increase knowledge of a sort that would be inscribed in a given space and time and constructed with reference to facts. Rather it is concerned exclusively with ensuring the conformity of operation procedures with standards developed by experts. This leads to positioning organization and practices beyond any control or supervision and excluding all critical analysis or debate of real work between those who prescribe and those who perform.

The second is the displacing of responsibility. If operation quality were understood to derive from collective organization of maintenance task management, then if incidents occurred, this would lead to evaluating working conditions and the way work is organized in such a way as to explain why the incidents happened, from both the viewpoints of operators on the ground and agents responsible for quality, and safety. However, when the "weak link" is the individual in charge of performing the prescribed tasks in compliance with quality assurance standards, then he becomes responsible for safety, and work conditions and organization are no longer taken into account.

We find here the same cognitive and didactic assumptions as those forefronted by Dassa and Maillard (1996) on the quality policy in the food industry. These assumptions inform the "company quality-safety" training process as a whole. The psycho-sociological assumption and the stigmatizing of subordinate task-performing personnel are especially strong because they concern "outside workers." Oversight of these workers' work takes the form of a moral judgment that is linked the relation of alienness (and therefore mistrust) obtaining between EDF agents and "outside" workers. For EDF agents, these workers are a threat to the agents' own safety. They do maintenance, a function that is essential to guaranteeing the reliability of nuclear installations, keeping in mind that "safety culture" designates human error to be what weakens that reliability. Since "outside"workers are the only ones able to operate to maintain installations, the threat of human error is obviously situated at just their level.

Lastly, it must be underlined that the entire "quality-safety" arrangement of the nuclear industry is based on company self-regulation. The bodies that develop certification standards for service companies and training programs for individuals are made up of owner-operators, and representatives of those service companies (GIIN, CEFRI). The public safety authority delegates to them the technical responsibility of nuclear installation safety. International standards are themselves fixed in the framework of nuclear industry expert commissions. The organization, conditions and practices of real work are absent from the processes by which these standards are developed. In the organizing of quality and safety for nuclear installations, questions on the relations between labor law rules and the interference between these rules and "company quality-safety" program standards are never raised. The nuclear industry has thus endowed itself with absolute power over how nuclear sites are organized and work conditions on them. Neither the state, as part of its public responsibilities, nor union organizations have any input into or control over the process.

CONCLUSION

The quest to be competitive can, in absolute terms, be considered contradictory with safety imperatives in that reducing production and maintenance costs at all costs implies an approach aimed at reducing financial resources. This is contradictory with the doctrine holding that everything must be done to guarantee safety. Still, in the social organization of maintenance operations, the nuclear owner-operator EDF has managed to circumvent this contradiction by seeming to reconcile the competitiveness objective with formal safety guarantees. Subcontracting allows for setting up flexible work organization and reducing payroll costs directly and indirectly. For 80% of maintenance, France's nuclear facilities authority is in no way held to institutionalized negotiations with union organizations around such crucial issues as working conditions, exposure to risk, and work schedules. With its economic legitimacy thus guaranteed, EDF has adopted a strategy that "insures" or "covers" it on matters of safety. On the one hand, it has endowed itself with a quality assurance policy that requires conformity of precedures as the very guarantee of quality. On the other hand, by making the organizational and human issues of the maintenance function a question of individual operators' responsibility, it has given itself a means of imputing incidents and accidents to the "human fallibility" of the person who signed off on a paper saying he performed the prescribed task. This protects those officially in charge of the nuclear industry from any suggestion that other components of organization or the decision makers who decided to organize operations as they did could be at fault. Can this conception of how to manage maintenance guarantee safety? Part III seeks to answer that question. Before doing so, however, it is important to examine how the second fundamental contradiction of the nuclear industry—a major issue for public health—is managed; i.e., the opposition between keeping nuclear plants safe and protecting nuclear maintenance workers from ionizing radiation.

CHAPTER 5

Organization of Radiation Protection for Nuclear Industry Workers: Rules, Norms, Practices

> Any distinction on the basis of origin, sex, family situation, state of health, handicap, lifestyle, political opinions, labor union activity, real or supposed membership or non-membership in an ethnic group, nation, race, or given religion, constitutes discrimination. (Art. 225-1 of the Code Pénal[1])

> The dispositions of the preceding article are not applicable if discrimination on the basis of state of health or handicap consists in refusing to hire, or in firing, a person who has been medically demonstrated to be unfit. (Art. 225-3 of the Code Pénal)

> Measures concerning safety, hygiene and health in the workplace must under no circumstance entail a financial burden for workers. (Art. L231-11 of the Code du Travail[2])

For nuclear owner-operators, plant maintenance organization runs up against another problematic contradiction, between the safety of facilities and radiation protection for workers. What are involved here are tasks—inspection, upkeep, cleaning, and repairing—that are indispensable for nuclear safety and that can only be performed by humans operating on the components of the nuclear production system. As we know, some of these components are highly radioactive, and radioactivity increases with the aging of the power plant. However, human intervention is only "socially acceptable" if the health of the workers involved is effectively protected.

[1] French penal code.
[2] French labor law.

Radiation protection poses first of all a problem of definition that immediately refers back to the difference between legal obligations and technical standards (Vacarie & Supiot, 1995).

According to French labor law (Code du Travail), protection against ionizing radiation falls under general prevention principles, specifically those that define the rules for protecting workers against the effects of toxic and carcinogenic products, the aim being to protect employees' health against any present or future deterioration due to their occupation (Code du Travail, Art. L230-2). These protective measures must also be consistent with the other legal provisions for ensuring safety and health in the workplace, as well as non-discrimination (Article L225-1 of the Code Pénal) and economic security (Code du Travail, Art. L231-11) (these last two articles are cited as epigraphs to this chapter).

Radiation protection as officially institutionalized in the nuclear industry refers less to any general principle of prevention than to the norms established for the industry by an international group of experts, the International Commission on Radiological Protection (ICRP). For the experts, these norms determine the "acceptable" risk level of exposure to ionizing radiation.

Nuclear operators can use two complementary strategies to comply with that risk level. One is the ALARA principle, which involves reducing radioactive discharge at the source; this implies taking technical measures or automating certain operations. It is costly, difficult to implement from the point of view of maintenance, and involves limitations inherent in the need for human intervention, although, according to the *Livre Blanc de la Radioprotection* [White book of radiation protection] (EDF, 1993), research has been initiated in this domain to develop "tele-controlled operation." The other strategy aims at reducing degree of individual exposure by breaking up and sharing out doses received so as to avoid going over the official mSv limits. But dividing up the dose this way leads to isolating personnel who have reached their dose limit and prohibiting them from entering the controlled zones (i.e., areas exposed to ionizing radiation). If those employees are part of the regular personnel of the nuclear plant operator, it would be very costly for the EDF employer to keep them out of the controlled zones for long periods and would pose insoluble or at very least difficult problems of personnel management and organization. As explained, the status of permanent personnel creates official obligations for the EDF employer concerning management of work hours and paid vacations, guaranteed employment, wages and the quality of working conditions, all of which have long been institutionally overseen by the labor unions.

As explained, subcontracting maintenance has permitted EDF to circumvent that contradiction by externalizing risk and delegating management of working conditions to the outside companies. Those companies have to show results with respect to quarterly and yearly dose allowance received by each individual DATR worker. EDF has made this mandatory and taken measures to ensure it is applied[3] without having itself to deal with how companies comply with the rule when tasks involving exposure to ionizing radiation are performed.

After a short presentation of how the International Commission on Radiological Protection developed its norms, we look at how they were incorporated into French labor law, then instrumentalized by EDF as a legal tool for managing job assignment by dose. In spite of being in contradiction with certain principles of the labor law and penal code, this tool has now been legitimized by the CNIL (*Commission Nationale de l'Informatique et des Libertés*, National commission for electronic information and personal freedoms), as we shall see further on.

ICRP RECOMMENDATIONS: THE KEYSTONE OF RADIATION PROTECTION

ICRP general principles are as follows (1977):

> Protection against ionizing radiation aims at protecting individuals, their descendants and humankind in general, while at the same time authorizing activities which may lead to exposure to radiation but are nevertheless indispensable. The detrimental effects that call for protection include somatic effects and hereditary effects. Effects due to radiation are said to be "somatic" when they affect an individual who has been directly exposed and "hereditary" when they affect his descendants. Those effects, which are likely to appear at a later time, regardless of their seriousness, are said to be "stochastic," and are considered to result from the dose received, to whatever degree. "Non-stochastic" effects are those for which the degree of seriousness varies by dose received and for which a threshold can therefore be established. . . .
>
> The aim of radioprotection should be to prevent non-stochastic detrimental effects and to keep the likelihood of stochastic effects down to levels judged acceptable. Another objective is to verify that practices implying exposure to radiation are justified.
>
> It is possible to prevent non-stochastic effects by setting limits of dose equivalents at sufficiently low levels that the ground limit cannot be reached, even if exposure extends over a lifetime or an entire occupational career. In order to limit stochastic effects, all justifiable exposure must be kept at the lowest possible levels that may be considered reasonable given economic and social factors, on condition that the appropriate dose equivalent limits are under no circumstances exceeded. (ICRP 26, 1977)

[3]EDF anticipated application of the norms in the 1996 European directive by establishing "warning thresholds" (between 15 and 18 mSv/year) determined in relation to the maximum value allowed in that directive (20 mSv) before this legislation was transposed into French law. EDF's company doctors were asked by directors of the various sites to help ensure that those thresholds were observed; as the figures show, this has lowered the number of "outside" workers exceeding the 20 mSv yearly limit.

Clarke, President of the ICRP (1996) adds:

> The real problem that scientists, the regulatory bodies and the public must solve does not concern risk level but risk acceptability. They must unite to determine what such acceptability is in different circumstances, both for workers and the public, in normal conditions as well as in case of an accident.

The International Commission on Radiation Protection is a non-governmental, not-for-profit organization composed of experts appointed by cooptation that has presided over the development of international radiation protection norms since 1959. Its expert recommendations are founded on two types of knowledge about the risks of exposure to ionizing radiation: biological and epidemiological information on pathogenic effects of ionizing radiation; cost/benefit economic analysis to determine what an "acceptable risk" level might be, i.e., the "acceptable" proportion of deaths due to cancer or genetic effects attributable to ionizing radiation in a given population.

The ICRP did not develop its standards with reference to the precautionary principle;[4] on the contrary, they were developed in accordance with scientific certainties acquired over time about the pathogenic effects of exposure on radiated populations, while the position adopted was a compromise between this scientific information and the economic imperatives of nuclear activities. The Commission's last recommendations drew contrasting reactions in France; on one hand from experts in favor of, or participating in, the nuclear industry (EDF/EPN 1996; Lallemand & Court, 1997), on the other from specialists of health in the workplace (Belbeoch, 1998; Huez, 1997; Zerbib, 1997). The French Academy of Sciences took the first position, in a report which has been challenged even by certain members of the research team who participated in the study (Académie des Sciences 1995).[5] The discussion centers on two main factors: rejection by nuclear industry and Academy of Sciences experts of ICRP's hypothesis that there is no pathogenic effects threshold for exposure to small doses, and the bases for determining what constitutes "acceptable" risk.

[4]"In a risk situation, a hypothesis that has not as yet been disproved should be considered temporarily valid, even if it has not been formally proved." G. J. Martin (1995) *Précaution et évolution du droit* (Paris: Dalloz): 299-306.

[5] The first Académie des Sciences report (1991) asserted that no new scientific facts justified lowering dose limits. A second report, published in October 1995 (Académie des Sciences, 1995) confirms this position, stating that "*there exist no indisputable recent scientific facts in favor of lowering the norms presently applicable to workers in France*" (p. 36). Several well-known contributors to the report spoke out against this position. Their position is published in an annex to the Academy's report. In a joint report published in 2005, the Académie des Sciences and the Académie de Médecine once again contest the validity of the linear non-threshold model (Académie de médecine, Académie des Sciences, 2005).

The Linear Non-Threshold Hypothesis

The Linear Non-Threshold Hypothesis holds that for a small dose, risk exists but tends to diminish with reduction of exposure dose. This international controversy has lasted for years and been the focus of numerous publications. Opponents of the model claim its validity has never been scientifically demonstrated. Consequently, nuclear industry operators feel that using it as a basis for rule design is a heavy and unnecessary economic burden (see especially Becker, 1998; Mossman, 1998). Researchers who support the linear model hypothesis justify it by quoting certain epidemiological studies (see Ashmore, Krewski, Zielinski, Jiang, Semenciw, & Band, 1998; Doll, 1997; Ritz, Morgenstern, Froines, & Bambi, 1999, in addition to those on which the ICRP based its recommendations). A group of international experts of the OECD's Agency for Nuclear Energy led by H. Métivier (IPSN-France) took a stand in this debate, presenting the various theses and listing the scientific "unknowns." They concluded:

> Given the present state of our knowledge and the precautionary principle, it seems justified to continue using the linear non-threshold dose-effect relation and the existing system of protection. (Commission on radiological protection and public health, 1998)

This controversy is further discussed in chapter 6. Here it is important to note that radiation protection exists and to understand how the level of "acceptable risk" was arrived at.

"Acceptable" Risk

The ICRP rejects the hypothesis that a danger threshold can be scientifically established that would allow us to understand exposure limits as a boundary between risk and no risk. It therefore characterizes the debate on effects of small radiation doses not in terms of scientific uncertainty—which it does not consider tenable given the results accumulated worldwide—but determining what "acceptable" risk might be.

In matters of exposure limits, the Commission's recommendations were developed by calculating a cost/benefit ratio, in which the expected benefit (the economic "necessity" of nuclear activities) justifies the human cost of that activity (expected number of deaths and genetic effects).

In 1977, the estimate was based on effect probability equivalence between risk of death due to occupational accidents in occupations known to offer high worker safety levels and risk of death attributable to ionizing radiation in an exposed population, a risk assessed on the basis of available epidemiological studies. The average annual mortality rate considered "acceptable" was 1 in 10,000. Extrapolating from this estimate, the ICRP recommended exposure limits that would be likely to decrease the dose received by an exposed worker in such a way that the probability of pathogenic effects appearing would not exceed 1 annual death from cancer for 10,000 workers exposed.

In 1990, after the results of epidemiological studies had revealed higher cancer rates in populations exposed to much lower dose levels than those previously accepted (Zerbib, 1997), ICRP experts changed their recommendations as well as their method for determining "acceptable" risk. They dropped the comparison with accident rate, admitting it was questionable to compare two types of phenomena as different from each other as workplace accidents and cancer and genetic risks associated with exposure to ionizing radiation. Based on an approach in price-of-a-human-life terms,[6] the level of "acceptable risk," i.e., the probability of dying from radiation-induced cancer during a career involving annual exposure to 20mSv over a period of 45 years, has been placed at 3.6%. Huez compared this with the risk level defined as acceptable in the case of asbestos, as estimated in an Institut National de la Santé et de la Recherche Médicale (INSERM) collective expert assessment report (1996), and found a risk level ten times lower than the one adopted by the ICRP for workers exposed to ionizing radiation.

Decrease in Life Expectancy Linked to the Effects of Ionizing Radiation

Clearly there is no danger threshold in the ICRP's new method for determining risk level. At the risk level it has considered "acceptable," decrease in life expectancy due to the collective dose of ionizing radiation received by French nuclear industry workers between 1977 and 1996 can be estimated at 880 years (see box below).

> According to calculations in ICRP 60, at this risk level a collective dose of 1 man-sievert causes a loss of life expectancy of 0.88 year. However, the cumulative collective dose received by exposed workers on EDF sites in France between 1977 and 1996, comes to ca. 1000 man-sievert, meaning loss of 880 years of life expectancy, keeping in mind that 80% of that collective dose is being taken by between 25,000 and 30,000 "outside" workers.

To evaluate the human cost, the reader is referred to the narrative at the end of chapter 3, which illustrates the fact that it does not involve merely material

[6]"Man-sievert price varies by country's development level. Thus, according to the ICRP, a man-seivert is worth $100,000 for an individual in a rich developed country, while the life of an individual from an underdeveloped country is worth only 1/33 that much" (Belbeoch, 1998).

cost but rather the death of contaminated persons and the loss and pain suffered by the loved ones they leave behind. In a society constantly on the lookout for medical means of postponing death, how is it that these deaths and this suffering, engendered by exposure to ionizing radiation—"in normal working conditions," have come to represent "acceptable" risk to the experts?

Prevention: Not a Matter of Limiting Exposure but Dividing Up the Dose

Under prevailing maintenance organization conditions, the human cost is invisible because the risk of radiation has been externalized and is borne only by "outside" workers. Organizing maintenance in this way causes occupational diseases connected to exposure to ionizing radiation to disappear from the employer's sphere of responsibility (as will be demonstrated in Part III). The adopted measures do not aim at controlling risk itself (i.e., the sources of radiation and radioactive contamination, and "dose discharge") but levels of radiation received by individuals. Contrary to the practice for other toxic products, prevention here does not consist in monitoring and reducing risk at the source by instating exposure maximums; radiation protection is instead and exclusively a means of precisely regulating doses individually received. This principle leaves the door open to selecting workers by dose. Exposure levels *per se* are only mentioned as a qualitative recommendation by the ICRP; they do not give any maximum figure. Indeed, this is the ALARA principle: a level "as low as reasonably achievable."

PRINCIPLES OF PROTECTING WORKERS AGAINST THE DANGERS OF IONIZING RADIATION IN FRENCH AND EUROPEAN UNION LEGISLATION

> Each member state shall ensure that the radiological monitoring system affords outside workers equivalent protection to that for workers employed on a permanent basis by the operator. (EEC directive of December 4, 1990)

Both European Union legislation and French regulations in affairs of protection against ionizing radiation have adopted compliance with ICRP recommended standards as a founding principle. The legal rules developed with this in mind are oriented not toward controlling exposure in the workplace, but rather supervising and monitoring individually received doses. The aim is to organize collection of quantitative data on dose using a system called dosimetrics, and to establish official means of guaranteeing that those limits are complied with.

The Norms

Directive 96/29/EURATOM, which replaced the 1980 directive, fixed the norms for protecting the health of the population and workers against dangers resulting from ionizing radiation in reference to those defined by the International Commission on Radiation Protection.

Basic Norms (EEC dir 96/29/ EURATOM):

- for workers, dose limits were lowered from 50 mSv/year to 100 mSv for 5 years, i.e. a yearly average of 20 mSv, with a yearly maximum of 50 mSv. In contrast to indications for other carcinogen substances, this norm does not indicate average amount of exposure per time unit (daily or monthly).
- for the population, dose limit was lowered to 1mSv/year.

The 1996 directive does not set any maximum threshold limit values (TLVs) for incorporation through respiratory and digestive tracts in case of internal contamination. Nor does it set any norms for uranium miners exposed to harm from radiation.

France waited until 1988 to transpose the 1980 basic norms directive into French law. The 1996 directive was transposed by Decree no. 2003-296 of March 31, 2003, adopted by the Ministry of Social Affairs.

Exposure maximums in effect in France since March 2003

A—concerning external exposure—*"exposure resulting from sources situated outside the human organism"*—commonly designated "radiation."

For workers, the limit on dose received over twelve consecutive months and evaluated by dosimetric techniques must not exceed 20 millisieverts (2 rems).[7] These limits may be exceeded in *"exceptional concerted exposure, in unusual working conditions, when other techniques cannot be applied,"* and in cases of "emergency exposure." Only Category A workers *"who have given their consent"* can operate in such cases, *"after receiving the appropriate authorization."*

In the earlier regulations it was specified that, once workers have undergone exceptional exposure, they may not be exposed to quarterly amounts of more than 1/10th of the yearly limit until their average exposure has once again fallen below the yearly maximum. Decree 91-963 specifies that these workers will benefit from the entire set of preventive and medical measures applying to category A workers, and that during the period stipulated above the employer must guarantee *"their employment along with a corresponding wage and a career plan that does not delay their promotion."*

[7] 1 rem = 10 millisieverts; 100 millirem = 1 millisievert.

For pregnant women occupationally exposed to ionizing radiation, the limit over the period separating declaration of pregnancy and delivery must be as low as reasonably achievable, and in no case exceed 1mSv. The limit for the public at large is set at 1 mSv/year, i.e., 20 times lower than for DATR workers.

B—Concerning *internal exposure—"exposure resulting from sources situated within the organism"*—commonly called "contamination," the "yearly limits of incorporation" are calculated separately for the various radio-elements, since radiotoxicity varies by nature of those elements. The elements are divided into three groups by degree of radiotoxicity: very strong, strong or moderate. Definition of the TLVs and cases exempt from the limits have been established by decree or ministerial order rather than in the framework of bills debated and passed in parliament.

Administrative texts of the Code du Travail

Decree 66-450 of 20/06/66 (modified by Decree 88-521 of 18/04/88) defines the general principles of protection against ionizing radiation. Decree 86-1103 of 2 October 1986 (modified by Decrees 91-963 of 19/9/91, 95-608 of 06/05/95 and 98-1186 of 24 December 1998) relative to protection of workers against the dangers of ionizing radiation, and Decree 75-306 of 28/04/75 (modified by Decrees 88662 of 6/5/88, 97-137 of 13/02/1997 and 98-1185 of 24 December 1998) relative to the protection of workers against the dangers of ionizing radiation in the INBs (basic nuclear facilities), set maximum exposure limits, as well as the administrative, technical and medical measures guaranteeing protection of employees assigned to work exposed to radiation and the specific modes for applying them inherent to nuclear installations. These texts were completed by the ministerial order of 31/07/91 creating a medical card for employees directly assigned to tasks exposed to radiation, by Decree 94-604 of 19 July 1994, creating the Office of Protection against Ionizing Radiation (OPRI), and by the Order of 17 October 1994, creating a "radiation protection authority" at the Ministry of Health. The Order of 12 May 1998, modifying that of 8 October 1990 already modified, establishes the list of tasks for which an employer cannot call on workers holding a limited contract or temporary workers, including "those operations that could lead to exposure to ionizing radiation if they are to be performed in zones where the dose output is liable to be higher than 2mSv/hour."[8]

[8]Since 2000 a set of decrees has been promulgated whose primary aim is to transpose the 1996 directive and centralize control of radioprotection within the French nuclear safety authority (see Preface).

Without offering any particular justification in prevention terms, the texts establish an important difference between maximum values for the population and workers assigned to operations involving radiation exposure. The latter clearly cannot be compared to a threshold between the presence or absence of danger. "Risk," even when considered "acceptable," is in contradiction with the founding principles of the Code du Travail for prevention of occupational risks, namely the rule that company owner is responsible for safety in the workplace. To get around this contradiction, Article 4 of Decree 86-1103 on owner's legal obligation to guarantee safety specifies as follows:

> Materials, procedures and organization of labor must be conceived in such a way that individual and collective occupational exposure is kept as low as reasonably possible under the limits prescribed by the present decree. To this end, exposed work posts must be regularly examined, with a frequency to be determined by exposure level.

As in ICRP recommendations, this provision is only qualitative, and includes no absolute obligation to reduce exposure. Yet the main rule of toxic risk prevention is to "limit dangerous substances." The fact that no average dose level for an 8-hour period is indicated contradicts that rule, since it permits intense though brief exposure of individuals within the limits of allowed quarterly or yearly amounts. This tends to transform dose limits into the "right" to get dosed or "radiation credits"—all still perfectly in keeping with regulations.

Finally, this regulation exempts employers from complying with the maximum value rules by applying "right to exceed "in" exceptional circumstances that are not defined in the text. The Force Ouvrière (FO) labor union contested the legality of Article 10 and subsequent articles of the October 2, 1986 decree for non-compliance with exposure limits, claiming they constituted a "violation of workplace safety conditions." In 1989 FO brought the case before the Conseil d'Etat, requesting annulment of the decree on those grounds. But FO withdrew its case—before the Conseil d'Etat had ruled—when the Ministry of Labor adopted the 1991 decree requiring employers to keep workers who had been exposed to radiation in the course of their work and to maintain their wage and career conditions. The question of whether exceeding the limits is legal has therefore never been settled.[9] This episode points to yet another feature of radiation protection regulations, one repeatedly contested by labor unions within the framework of the *Conseil Supérieur de la Prévention des Risques Professionnels* (Higher council for the prevention of occupational risks) of the Ministry of Labor: the imprecision of administrative texts. Conditions of application consistent with other legal and administrative Code du Travail provisions have still not been specified, either by ministerial order or official application instructions.

[9]Appeal filed with the litigation department of the Conseil d'Etat by the J. Guinard on behalf of the FO labor union, 1989.

Regulation Dosimetric Monitoring

Supervising dose received is crucial for ensuring that norms are being complied with. The regulations therefore specify exactly how doses are to be recorded, which service should handle the data, and what the status of the information collected on individuals is. This is what is called administrative dosimetrics, defined in the texts as a section of "measures of a medical nature." It is compulsory, and formally designed to allow for personal follow-up of employees, to protect their health by ensuring that dose limits are not being exceeded and to keep a life-long record of doses received during individual's occupational career. The technique is to record doses received on film, which is then developed by an independent public service unconnected to and independent of the owner-operators. The results of this administrative dosimetric monitoring, handled first by the SCPRI, then by the OPRI and since June 2004 by the SISERI[10], are confidential; they are an integral part of a worker's medical record and as such covered by medical privacy. The information is given to the occupational doctor in charge, who in turn communicates it to the employee.

Medical Surveillance

European Union and French regulations are characterized by precise rules for individual medical surveillance. "Measures of a medical nature" structure individual surveillance of DATR workers, in collaboration with the public service charged with the dosimetric follow-up, today the SISERI. Individual surveillance consists in a bi-annual medical check-up. The exam is not aimed at determining a person's fitness for a specific job whose potential risks for a given employee's health are to be assessed by the doctor, but rather to decide whether the employee may or may not be classified Category A (i.e., directly assignable "in normal working conditions" to tasks exposed to radiation, whence the acronym DATR: Directement Affecté aux Travaux sous Rayonnements). An affirmative response indicates "absence of any medical counterindication for tasks exposed to radiation." This clause contradicts the legal rule according to which the occupational doctor's duty includes protecting employees' health in the workplace in accordance with general safety and prevention principles.[11] In the present case, the rules allow a worker medically confirmed as fit to perform tasks exposing him

[10] Système d'Information de la Surveillance de l'Exposition aux Rayonnements Ionisants, http://www.siseri.com/siseri

[11] Article R241-41 of the Code du Travail stipulates the duties of an occupational doctor as follows:

- to improve living and working conditions in the company;
- to adapt work posts, techniques and work pace to human physiology;
- to protect employees against all harm, in particular against the risk of accidents in the workplace and risks attached to use of dangerous products;
- to maintain the general level of hygiene on the premises;
- to supervise hygiene in restaurant services;
- to educate on health problem prevention in connection with occupational activities in the firm.

To fulfill these duties, occupational health doctors are to carry out medical examinations and take actions that will positively impact on the workplace.

or her to cancer risks as long as they comply with exposure norms—as if those norms determined a danger threshold beneath which there were no threat to health.

Several other provisions (medical measures presented in the box below) have been adopted over the past ten years concerning medical record, individual follow-up and certification of any medical services regularly checking DATR workers. However, these texts have nothing to say about the company doctor's actions to improve working conditions and reduce exposure in the particular work context of INBs.

Medical measures

Provisions in Decree 75-306 of 28/04/75, modified by Decree 88-662 of 6/05/88 and Decree 2003-296 of 31/03/03, set the calendar for medical visits and fitness exams, and determine ordinary and special medical record content for Category-A workers, i.e., those directly assigned in normal working conditions to tasks exposed to radiation.

Category-A workers must have a medical check-up every six months (including a general clinical examination, to which is added one or several special exams depending on type of exposure). The check-up is used to determine whether they are fit or not to perform tasks exposed to ionizing radiation.

A special medical file includes

- "a form indicating working conditions of the exposed worker, which must mention type of task, characteristics of radiation sources and nature of radiation, length of exposure periods and other job-related risks or harm of physical or chemical origin. The form is filled out by the employer, the radiation protection service or personnel under employer's responsibility, as stipulated by Article 8 of the present decree, and must be signed by the worker. It is to be regularly updated;
- "an exposure form mentioning dates and exam results for individual exposure and exposure duration;
- dates and results of medical examinations."

The special record must be kept throughout worker's lifetime and at least thirty years after he has ceased DATR activity.

The role of the Service Central de Protection Contre les Rayonnements Ionisants (SCPRI) was specified in detail in the 1988 Decree. This service was replaced by the IRSN, in charge of developing individual dosimetric films and recording the results, which it has the exclusive right to keep and use. However, certain private laboratories have obtained a license to perform the same functions. This is also the case for owner-operators of basic nuclear facilities. The ministerial order of 31/07/91 instituted a medical follow-up card, distributed to Category-A DATR workers by the occupational doctor andregistered at the national level with the OPRI. The card was in two parts: one for the worker, the other addressed by the doctor to the OPRI for each card delivered. The information written in by the doctor was then computerized; it included consecutive medical visit dates, doctor's identity, and mention of whatever restrictions there might be concerning DATR worker's fitness and whether or not he was being followed for biotoxicity. The card was valid for a maximum of three years. It did not contain dosimetric results. The Order of 28 August 1991 (abrogating the preceding text bearing on the same subject) gives specific recommendations to occupational doctors in charge of medical surveillance of workers exposed to ionizing radiation. It defines surveillance modes as follows: doctors are to refer to the SCPRI (later OPRI) to establish the list of any previous exposures, and consult the medical file before making an unconditional decision about Category-A fitness. If precise data on exposure are not available, the Order indicates that the doctor must give an estimate; if for a given worker the amount of exposure seems too close to the maximum values, there is nothing to prevent the doctor from declaring him fit anyway, with conditions, and for less than six months, implying that the missing dosimetric monitoring will be supplied in the interval. If the dosimetric monitoring results cannot be found, the worker is said to have reached the dose equivalent corresponding to exposure limits for the given period.

Lastly, Order 97-137 of 13 February 1997 abrogated the provisions stipulated in Decree 92-158 of 20 February 1992 on intervention by outside companies according to which "the user company will be responsible to the outside company for the complementary examinations required by the nature and duration of the tasks carried out in that company by outside company employees" (Code du Travail, Art R237-19). It transferred that responsibility to the medical services of the outside company, meanwhile creating the requirement of special certification for outside company medical services and special training for doctors in charge of outside company DATR employees working in INBs. The minimum time allotted for medically following these employees' was thereby increased to more than what is usually allotted in industrial medicine.

The order of December 30, 2004 on personal medical follow-up cards and individual dosimetric information for workers exposed to ionizing radiation changed the content and modes for being granted a personal medical follow-up card, stipulating as follows: "The card number delivered by the IRSN covers all periods of the worker's occupational life that entail exposure to ionizing radiation."

Protection of Outside Employees Exposed to Ionizing Radiation during Operations in Controlled Zones

The 1990 European Union Directive

In response to the fact that tasks exposing workers to ionizing radiation were being subcontracted by the nuclear industry throughout Europe,[12] the EEC passed the 4/12/90 Directive 90/641/EURATOM concerning what is called operation protection of outside employees exposed to ionizing radiation during operations in controlled zones. The basic principle contained in this text is that "every system of anti-radiation surveillance of outside employees shall, by means of common measures, guarantee protection [for outside employees] that is equivalent to that for permanent owner-operator employees." The very fact that this principle had to be written into the text reveals its ambivalent character, since it bases equal right to protection against ionizing radiation not on measures that would control the risk at its source but on the efficiency of individual dosimetric control.

The Directive also enumerates the respective obligations incumbent on European states, nuclear owner-operators, and outside companies (see box on next page).

In 1997, the EEC addressed a formal order to the French state to integrate the 1990 directive into French law. For certain provisions this was only done in late 1999. Operational dosimetric monitoring in its current form—i.e., as a system of managing employment by dose—was legalized through administrative channels. But as will be discussed further on, the directive itself—before being integrated into French law—was used as a legal basis in the joint practical organization of individual dosimetric monitoring supervision, termed "operational" by owner operators and outside companies operating in the nuclear industry.

French Legislation on Health and Safety of DATR Outside Employees

The modes of medical surveillance contained in the decrees on radiation protection of outside employees working in INBs may be seen as a kind of recognition of the different statuses of owner-operator employees and outside company workers. However, concerning the responsibility for overall coordination of the preventive measures adopted by EDF and all executive managers of

[12] For 1991 the OECD published percentages of total collective dose received by outside personnel in nuclear power plants of type **REP** in OECD countries. These ranged from 67% in Switzerland to 97% in Japan. France and the European Union as a whole were both at 82%. Also published was proportion of total dose received during unit shutdowns. It ranged from 73% in the Netherlands to 97% in Japan; 83% in France, 82% for the European Union. Figures cited in Birraux, 1993.

State obligations: "Each member-state ensures that the radiation surveillance system guarantees outside workers protection equivalent to that granted permanent employees of the owner-operator. Until such time as a uniform system of radiation protection for outside workers, such as an electronic network, can be set up at the EEC level, this is to be done either through a centralized national network or an individual radiation surveillance document delivered to each outside employee."

Owner-operator's obligations: "The owner-operator of a controlled zone in which outside employees operate is responsible, either directly or by contract, for the operational aspects of radiation protection directly related to the nature of the controlled zone." Owner-operator must make sure that employees are certified medically fit, that they have received basic training in radiation protection, that they have individual equipment, benefit from individual and operational surveillance of dosimetric monitoring results, that dose limits are complied with, and that at the end of an operation, radiological elements have been recorded.

Outside company's obligations: The outside company shall ensure that radiation protection of its employees is guaranteed *"either directly or by way of contractual agreements with the owner-operator" while* ensuring compliance with general principles and dose limits, providing training and information on radiation protection as well as medical surveillance and assessment of individual exposure.

outside companies operating in EDF plants, the important reference is Decree 92-158 of 20 February 1992, which "stands as a supplement to the Code du Travail and prescribes particular conditions for hygiene and safety applicable to tasks performed by an outside company." The text stipulates: "the present general coordination aims to prevent the dangers linked to interference between activities, installations and equipment of the various companies present in the same workplace."

As pointed out in chapter 4, this decree tends to relieve owner-operator of responsibility for the working conditions of "outside" workers. It states, for example, that before an outside company can start working on a user company site, a prevention plan must be drawn up, including description of work to be performed, instructions for employees, and measures to be followed. The list of work posts occupied by employees likely to come under special medical surveillance must figure in the plan. The list does not mention the names of employees at these posts. Nor is there any mention in the plan of practical consequences that implicate owner-operator doctor in outside worker health protection since special medical surveillance of outside workers is itself sub-contracted to outside companies' industrial medicine services by exemption from regulations of the February 1992 decree (Order 97-137 of 13 February 1997, see "Medical measures" box).

French Legislation on Health and Safety of DATR Temporary Employees

The law of 12 July 1990 prohibits using temporary workers to carry out the dangerous tasks listed in the ministerial order of 8 October 1990. Ionizing radiation did not figure on the list until adoption of the Order of 12 May 1998 (cf. *Journal Officiel* of 28 May 1998). This order prohibits resorting to temporary workers or limited contracts to fill work posts involving high exposure. However, the level in question is so high that it concerns only approximately 10% of exposed maintenance posts in the nuclear industry, posts to which temporary workers are in fact presently assigned, "in ordinary working conditions" (Huez, *Le Monde,* 19 Feb. 1998). Moreover, employees with what is called a contingent work contract— of (single projects, unfixed duration)—i.e., those whose contracts are the easiest for the employer to terminate—are not affected by the decree.[13] Independent workers are not concerned by the measure either.

Other than this recent text and those mentioned in chapter 4 defining modalities for using temporary employment, the only particular health-in-the-workplace provisions for temporary workers have to do with medical surveillance (Decree of 23 July 1991).

Clearly the only areas in which "outside" DATR workers' rights are in any way legislatively equivalent to those of owner-operator permanent employees are medical and dosimetric surveillance. European legislation and French regulations do not include any specific or clear provisions for reducing the structural inequality of dose levels received by permanent and temporary "outside" employees, on the one hand, in-house personnel on the other. Nor is there any provision preventing selection and either partial or total lay-off of "outside" employees who reach their dose limit, or compensation for any economic or occupational losses caused by such selection.

[13] With regard to contingent work contracts, a letter addressed to directors of French regions and *departements* from the director of labor relations at the labor ministry indicates that permanent contracts are not affected by the decision of 8 October 1990. Nevertheless, Marimbert adds with regard to jobs exposing workers to asbestos (also on the list of jobs covered by this order) that "increasing use of contingent work contracts to fill labor requirements involving removal of asbestos could be considered an abuse and turn out to be detrimental to the objectives of the asbestos policy, which are to protect employees' health" (Marimbert, letter of 13 March 1997). The same is of course true of jobs involving exposure to ionizing radiation.

IN PRACTICE: WEAKENED DOSIMETRIC FOLLOW-UP AND MANAGING JOB ASSIGNMENTS BY DOSE. THE ROLE OF THE CNIL (National Commission on Electronic Information and Personal Freedoms)

The state has been slow to implement the texts on protection of employees against ionizing radiation. For decades there were no ministerial orders specifying how existing decrees on protection against ionizing radiation were to be applied. As mentioned, in order for the French legislature to integrate the 1990 European directive, it took an official warning and order from the EEC—eight years after the European Community adopted that directive. Moreover, the public system for administrative collection of dosimetric monitoring results functions so poorly that its authority has been discredited. An overview of the situation by the French MP Le Déaut brought to light resource dispersion and the inefficiency of the public radiation protection system (Le Déaut, 1998). During those years, owner-operators were undertaking a series of steps to obtain the necessary authorizations from the CNIL for legalizing the "operational" computerized system for managing dosimetrics. This system has became a fundamental means of submitting "outside" workers who are reaching their dose limit to a selection process of job assignment by dose.

The Emergence of "Operational" Dosimetrics

At this stage, it is important to return to the theoretical definition of the two sorts of existing dosimetric monitoring: administrative and operational.

Administrative dosimetrics involves precise legal rules (see above), which have been included in the regulations. It refers to administrative surveillance of employees and the requirement that they deliver dosimetric film to the OPRI for development. The results remain confidential; the medical privacy law applies: owner-operators and their doctors do not have access to information on employees from outside companies.

Operational dosimetrics, on the other hand, means surveillance of exposure received at worker's post *during the operations*. Since 24 December 1998, this type of dosimetrics is regulated by a decree that went into effect in late 1999, which is understood to transpose the 1990 European directive on protection of outside employees working in an INB. The owner-operators themselves therefore set up the "operational" dosimetrics surveillance system, without any specific administrative or legislative oversight. The objective of operational dosimetrics is to implement the ALARA system, i.e. reduce dose at the work post to the lowest level possible. Operational dosimetric measures are recorded by means of a different technique than the one used by administrative dosimetrics, an *electronic "saphymo"* that allows for immediate access to dose received. Operational dosimetrics, also called "electronic dosimetrics," are designated in the European Union directive a compulsory practice for the owner-operator.

As actually applied, these two ways of collecting dosimetric monitoring information have in common the fact that they are individual, name-specific records. They are differentiated not only in terms of techniques applied but also objectives and the legal status of the information collected.

In 1990, at a meeting of the Société Française de Radioprotection, Mayoux recalled the objectives of dosimetric surveillance, insisting on the difference between individual administrative dosimetrics and operational dosimetrics.

OPERATIONAL DOSIMETRICS

(Quoted from Mayoux, 1990)

"The term 'operational dosimetrics' does not appear in regulations; however, the regulations provide for several situations where employer is required to carry out dosimetric monitoring that may be qualified as operational.

These are

- a periodic analysis of work posts, so as to keep individual and collective exposure 'as low as reasonably possible';
- the need to calculate and check dose level that workers will be exposed to and determine maximum exposure time for each worker *before* execution of exceptional tasks, particularly repair and upkeep tasks.

In complex installations, individual dosimetric monitoring and "regulation" atmosphere checks may prove insufficient to meet these imperatives, especially since film results can only be known after a time lapse, as they are read only once a month.

With operational dosimetrics, dose received can be read immediately, so this information can be used instantly. This makes it possible to determine extremely quickly the exposure level corresponding to a given work post or operation. It is thus a fundamental element for optimizing dose received by a given worker. Operational dosimetrics can also be a complement, in the sense that it allows for using dosimeters that are better adapted to the nature of certain types of radiation and operations. The results must be made available to the occupational doctor; the rules stipulate that in the case of exceptional exposure during an operation, whether scheduled or emergency, the worker must have a means for measuring individual dose that is adapted to the particular conditions of the exposure.

Operational dosimetrics is not normally covered by medical privacy, but it seems legitimate to ask if individualized operational dosimetrics would not be a good idea, alongside legal dosimetrics, as it would allow parallel dosimetric surveillance of workers."

The stated reason for recording individual dosimetric results is reducing dose emitted at worker's post, which requires that workers be aware of the amount they are receiving as they carry out their operations at that post. Given that maintenance operations involve high exposure levels (cf. workers' narratives in Part I), the major implicit objective for the owner-operator is much more likely to be sharing out exposure among a large number of "outside" workers, while making sure they do not exceed the authorized limits—whence its interest in having information about the annual doses accumulated by each identified worker. However, the legal rules surrounding the collecting of such information involve decrees stipulating official dosimetrics procedures. These are strict (despite the fact that orders specifying how they are to be applied have yet to be drafted): this is information about workers' health, and therefore covered by medical privacy.

In objecting to the use of dosimetric surveillance systems that fall outside the realm of industrial medicine, Chanteur of the SCPRI recalled the possible "counter-effects" of "illegitimate transfer of medical competence" to owner-operator or an outside employer:

> Such illegitimate transfer of medical competence has countereffects, including the following:
>
> - An incentive to let workers receive the highest dose limit, i.e. as much as the rules permit. Those limits will then become a value to attain in order to get the most out of a worker, which goes against the fundamental principle of optimization;
> - An incentive to exclude and even lay off workers whose cumulative dose makes them unemployable, with all the labor and social consequences implied for these workers and the temptation for them to destroy the compromising record (Chanteur, 1990).

The practices used show the perspicacity of this analysis. Within three years of the decree, nuclear industrialists found a means—legitimized by the CNIL—of using operational dosimetrics to manage employment by dose, without even having to resort to lay-offs. That method is subcontracting and using temporary labor.

Medical-Administrative "Disorder" in the Radiation Protection of Outside Company Employees

In 1988, when EDF began regularly subcontracting outside firms to perform its maintenance jobs, certain CHS-CTs in nuclear power plants and some occcupational doctors began asking hard questions and expressing strong criticism about the poor functioning of SCPRI-managed medical and dosimetric surveillance of "outside" DATR workers operating in controlled zones. These critiques led the Ministry of Labor in conjunction with INSERM to commission a study of the concrete modalities of medical and regulation surveillance of outside DATR workers, and malfunctioning of the radiation protection system (Thébaud-Mony et al., 1992). Against a backdrop of increasingly precarious occupational itineraries for the 87 employees who participated in the study (frequent changes of company, intense geographic mobility, temporary work alternating with short-term contracts, intermittent activities), the results of the study revealed the following: the very notion of medical follow-up seems to have disappeared, replaced by metronomic observance of biannual check-ups, themselves transformed into "fitness rituals." The 87 employees' experiences of occupational medicine show that during the four years of the study, 103 different doctors spread out over all regions of France had intervened.

Dual dosimetric surveillance:

- On one hand, very poor management by the SCPRI of the regulation dosimetric monitoring of outside company workers. Information concerning the films arrived late and was subject to caution. Indeed, a number of unidentified dosimetric films were distributed but never developed, so that it was impossible to know if they had been used and by whom. Furthermore, absence of a data bank makes it unlikely that any reconstitution of a worker's dosimetric record could ever be used to evaluate the reality of exposure in case of an occupational disease. This also makes it impossible to arrange for epidemiological surveillance of the population submitted to the greatest amount of occupational radiation in the INBs. On the basis of the data gathered during this study, for 80% of employees in the sample it was impossible to retrace a dosimetric career longer than one year.
- At the other extreme, the EDF owner-operator's efficient management of electronic dosimetrics—in place of the administrative variety—on its own sites. Instead of being used to control exposure at the work post, regulation dosimetric monitoring has become a tool for measuring individual dose levels. The authors of the study stress that operational dosimetrics, freed from all administrative oversight and in the context of intensely precarious employment as mentioned above, tends in fact to be used as a technical criterion for selecting employees (medical fitness). Given the modalities by which doctors follow outside employees of the nuclear industry—reduced to a bi-annual "fitness ritual"—it is fair to say that occupational medicine has become instrumental in conferring medical legitimacy on management of employment by radiation dose.

The results of other studies carried out with different methods in other parts of France all concord (Forest & Zerbib, 1991; Pille, 1991). All show structural inequality between doses received by outside employees in comparison to in-house employees.

In 1991 the situation was such that the European Commission intervened. In a letter of 24 May 1991 addressed to EDF, the Commission demanded an explanation for the discrepancies between its figures and those of the SCPRI on yearly dose collectively received by outside company employees. The discrepancy: in 1987 for the entire nuclear industry, the SCPRI declared the collective dose to be the equivalent of 26 man-sievert for 7585 outside workers, whereas for its power plants alone EDF declared a collective dose of 67 man-sievert for an "estimated" population of 20000 sub-contracted workers—a difference of 300% per year, according to the EEC. The accusation was made public in a newspaper article presenting an interview with managers of EDF's radiation protection system who were concerned about the situation.[14]

Between 1991 and 1995, the medical-administrative "disorder" brought about a flurry of activity in two advisory bodies, one a parliamentary committee (public hearings, and report by Birraux on behalf of the *Office Parlementaire d'Evaluation des Choix Technologiques* [Parliamentary office for the evaluation of technological choices], the other the joint administrative taskforce of the *Conseil Supérieur de Sûreté et d'Information Nucléaire* (High council of safety and information on nuclear activities). These bodies made two rulings, one in January 1992, the other in March 1995, stressing the responsibilities of the state and the owner operator, with regard not to the legal rules of the Code du Travail on protection of DATR workers' health—of which radiation protection is but one aspect—but rather in relation to "the cornerstone of the radiation protection policy: controlling and limiting dose," to use Claude Birraux's expression. The malfunctioning denounced therefore concerned only two aspects of the issue. The first of these was absolute compliance with maximum dose values. The understanding of these values was constructed from the concept of "radiation protection," which tends to transform those limits into a danger threshold. The second was the need to reorganize and regain control of dosimetric surveillance of outside workers, and to ensure *protection equivalent* to that for in-house employees. Here the state was called upon to guarantee not only dose control but also the transparency of such control. In the new regulations that came out of this intense advisory activity, the SCPRI was abolished—its credibility had been severely damaged—and a new body, the OPRI, put in charge of reorganizing official dosimetrics and strengthening existing medical surveillance measures.

But two crucial aspects of the problem of radio protection for outside employees in connection with health in the workplace, questions raised at the *Conseil Supérieur de Sûreté et d'Information Nucléaire* (CSSIN) by union representatives, were not dealt with: *the structural inequality of doses received* and the harm this causes to "outside" employees doing maintenance jobs; and the fact that workers were increasingly being *selected according to maximum exposure limits*—the equivalent of a "radiation credit," in the words of Claude Maugin, FO union delegate to the CCSIN.

[14] *Libération*, 23 October 1991.

The CNIL Legalizes Management of Hiring and Job Assignment by Dose

In addition to recording official administrative dosimetrics on film, EDF has long been using electronic dosimetrics for information on doses received during an operation, both individually and collectively at the time of radiation emission. Real-time recording allows DATR employees to be immediately informed of their exposure. Up until 1992, they noted doses received as established by electronic dosimetrics in a booklet, and from the film when available. The booklets had no legal status; they were known as the "green book" and belonged to the employee. The idea originated with the Groupement Intersyndical des Industries du Nucléaire (GIIN; United nuclear industry unions).[15] The access booklet needed to enter the power plants has now replaced it.

In 1989, EDF decided to computerize electronic dosimetrics results for all DATR employees, in-house and subcontracted, in all French nuclear plants. These moves, parallel to the ones by the GIIN under the egis of the CNIL, led the CNIL to legalize computerization of cumulative nominative dosimetric data, thereby making it available to all employers (EDF and outside companies). I determined the chronology of these developments by consulting the archives and compiling all requests for rulings filed with the CNIL.[16] Legalization was thus obtained by stages.

December 1989: File P2

EDF requested a ruling from the CNIL on automated processing of name-specific dose information in order to establish records for "technical and administrative human resources management" (file P2) for both EDF agents and the "outside labor" operating on its various sites. Among other information on these workers were the individual data from operational dosimetrics. The stated objective here was very broad: "to improve technical and administrative management of human resources and facilitate maintenance management." The CNIL requested no further information and did not look into the reasons or question the legitimacy of creating a file containing name-specific information on outside company employees, i.e. employees not linked to EDF by contract. It did not require EDF to justify the status of dosimetric information with respect to the existing legislation. Consent was given three months after the request was filed (March, 1990).

[15] The GIIN was created in 1959 by 12 professional federations. In 2000, it covered 190 companies operating in the French nuclear industry and employing 20,000 DATR workers.

[16] The different steps are presented in detail in a research report done for the Fonds d'Intervention en Santé publique in December 1995 (Thébaud-Mony, 1995).

November 1991: DOSINAT

In its next request to the CNIL, EDF asked that the previously authorized P2 File dosimetric module be modified so as to be able to keep the name-specific data from operational dosimetrics for five years instead of the fourteen months provided for in the P2 file. It was also in 1991 that the ICRP issued new recommendations, setting the norm of maximum exposure at 100mSv over a 5-year period (CIPR60). This was done shortly after the EEC had adopted the December 4, 1990, directive on protection of outside workers operating in nuclear installations. EDF justified its request to the CNIL on the grounds of these two texts, neither of which had been integrated into French law.

On the other hand, the owner-operator was silent about the link-up of files from the various power plants that was necessary to ensure computerized processing of the doses accumulated by itinerant workers forced to circulate from one site to another on the subcontracting market. Article 19 of the "electronic information and personal freedoms" law of January 6, 1978 requires any company requesting to set up or modify a computerized system to "specify the linking, interconnecting and all other forms of data link-up as well as any communicating of this data to third parties."

The owner-operator also said nothing about all the legal debates and interrogations on use and privacy of dosimetric data, and about the warnings, particularly from SCPRI experts, on the danger that operational dosimetrics could imperceptibly become a discriminatory system of managing employment by radiation dose. The CNIL asked no questions and authorized the modification one month after receiving the request.

Parallel to the request for modification filed with the CNIL, an EDF-GIIN protocol[17] provided for creation by the GIIN of an "intermediate file" that would contain all the dosimetric records of outside employees, to be regularly updated (twice weekly, for example) either by transferring the information contained in company files (first EDF's files, later those of other nuclear companies if contacts with them were successful) or by including information provided by outside employers. This file would be compatible with EDF's. Owner-operator and previous outside company employer would thus have access to the name-specific information in the file, which would be managed by the GIIN (or a subcontracted company). EDF pledged "to make the information needed to update outside personnel dosimetric records available for subcontractors' intermediate file." The protocol mentions the requirement of getting CNIL approval for setting up this integrated system of individual name-specific surveillance of outside employees' dosimetrics.

[17] Protocole signed by the DPT director (represented by M. Carlier) and the GIIN (represented by M. Mathias) on November 20, 1991.

May 1992: Access Booklet for Entrance to the Power Plant

At this stage, it was the GIIN through its commercial subsidiary, PROMINF (*Promotion de l'Industrie Nucléaire Française*), rather than the owner-operator that filed a standard declaration of intention, in this case intention to process name-specific information electronically in connection with "managing the booklet for access to power plants, including identification of operating personnel and their companies and an ID number." In addition to identifying information, this record on outside company employees includes individual operational dosimetric results that may be kept for a maximum of five years. Nothing is said of the methods used to collect the operational dosimetrics. The request states that the aim of the access booklet is to provide the owner-operator EDF with "a simple means of verifying that outside personnel comply with the conditions required for operating on a nuclear site, especially in matters of radiation protection."

For the workers, whether they have a right to consult the file containing information about them depends on their employer.

The legal status of the two structures (GIIN, PROMINF) is not specified in the documents I consulted at the CNIL, nor do the documents explain what makes it legitimate for those organizations to manage a name-specific file of employees of different companies with which they have no contract relation.

After expressing disagreement with the mention of nationality (which the GIIN later eliminated) and length of time the data may be kept—which it ended up approving—the CNIL gave its approval to the declaration.

September 1992: Linking Up EDF and GIIN Records

EDF asked the CNIL to approve another change in the P2 file, this time for giving the outside companies access to information on operational dosimetrics. Once again, the owner-operator justified its argument by quoting the December 1990 directive. The CNIL asked for the text of the directive and the Act by which it had been transposed into national law. EDF's answer was to assert that, according to a decision of the European Court of Justice, the directive applied even if it had not been transposed. The CNIL granted permission.

August 1993: DOSIMO

The GIIN took over the procedure initiated by EDF. It requested the CNIL to rule on computerized processing of "combined management of doses received by outside company employees operating under exposure to ionizing radiation in the French nuclear industry (DOSIMO)." The point was to link up the files of the various owner-operators (EDF, CEA, COGEMA, the Army) with the file that PROMINF was managing for the GIIN containing access booklet data, and for the information to be made available exclusively to the owner-operator, the subcontracted companies, and persons in charge of radiation protection; i.e., neither the employees themselves nor the CHS-CT or company doctors were to be allowed access to the information. CNIL authorities were surprised by this request, but the GIIN's subsidiary, PROMINF, responded by quoting the European Union directive of December 1990, justifying its request by citing the responsibility incumbent upon the state and the fact that it was up to the owner-operator and outside company employers to ensure dosimetric surveillance of outside employees. Once again, the CNIL gave a favorable ruling.

That a worker has nearly reached the maximum dose can be deduced from the name-specific information contained in the interconnected electronic processing system. This practice no longer has anything to do with the official objective of operational dosimetrics; i.e., the principle of reducing on-the-job dose received in the course of operations. Having reached maximum dose is then cited to the worker as the reason for refusing him access to a nuclear site, and no medical mediation or recourse of any sort is possible against that form of implicit selection (or discrimination). The fact that the file is in owner-operator hands means that owner-operator can refuse entrance to workers with a high cumulative dosimetric level.[18] Accessing that information allows employers to organize their selection of DATR workers before they arrive. No particular item in the management of these files counteracts the fact that they can be selected through computerized information on dose. These are the practices workers are referring to when they speak of reaching the "peg"—i.e., maximum dose (see Part I). Since EDF is not the direct employer of the workers involved, it does not have to deal with any consequences of a refusal of site access. Those consequences are what follow from selection on the basis of past exposure, which tends to exclude from employment workers who have used up their "radiation credits."

These practices go against at least three specific legal rules. Article 2 of the electronic information and personal freedoms aw of 6 January 1978 states that "no administrative or private-sector decision implying the evaluation of human behavior can be based on computer-processed information that produces a profile or defines the personality of the individual concerned." The law of 12 July 1990 forbids all discrimination in the workplace connected to a job applicant's

[18] In the Saint Laurent des Eaux plant, for example, an internal memo (no. 270898) regulates entry to the site by outside workers by 1) indicating "warning thresholds" that prohibit "outside" workers from going through the plant door (the daily limit is set at 2 mSv, monthly limit at 4 mSv, 3-month limit at 12 mSv, 12-month limit at 18 mSv); 2) creating a card to be filled out by an outside worker who has exceeded the warning threshold, called "*identification of worker over the warning threshold*" determining "*projected dosimetrics,*" and requiring employee to sign the following statement: "*In signing this document, I swear to have given all my dosimetric information and that I am aware of the impact the new thresholds will have on my future activity.*" Some occupational doctors objected to this practice, recalling that workers are not responsible for doses absorbed.

state of health, except in the case of medical unfitness indicated by the company doctor. Lastly, the Law of 31 December 1991 (Art L231-11 of the Code du Travail) stipulates:"the measures concerning safety, hygiene and health in the workplace may in no case entail any financial burden for employees." But even permanent employees, "put out to pasture" because they have reached the maximum permitted dose, are forced into partial unemployment leading to major income reduction. However, when outside employers themselves, owner-operator at the entrance to their plant, or the persons hired by one or the other to handle radioprotection consult the file electronically, this leaves no trace. Workers subjected to premature termination of their contract or operation or who are not hired or laid off because they have used up their "radiation credits" thus cannot prove that the record was consulted illicitly or bring charges. This is only possible before a court or the CNIL in case of blatant discrimination. The system is virtual; use of it is therefore invisible.

December 1996: The Protocol Signed by the OPRI, Owner-Operators and the GIIN

The penultimate move in this strategy was a "protocol for a national external dosimetrics network on worker's operating in nuclear installations and exposed to ionizing radiation," signed in December 1996 by the OPRI and nuclear industry owner-operators and entrepreneurs. Referring once again to the European directive of December 4, 1990 which had still not been integrated into French law, it proposed to assign OPRI to manage the interconnected name-specific operational dosimetrics data bank for France's nuclear installations. OPRI was equipped by the GIIN with an active terminal from which to access the system without any of the legal constraints for official administrative dosimetrics. By signing the protocol, the OPRI as a state organization confirmed the legitimate nature of the name-specific system of information on cumulative, individual dosimetrics set up and controlled by nuclear entrepreneurs and industrialists, thus striking out in a single line of ink all the legal rules applying to official dosimetrics.

It should be stressed that the OPRI had already received the green light for computerizing official dosimetrics from the CNIL several months before this protocol was signed (the signing took place in 1997), though the administrative act permitting its implementation had not yet been formalized.

December 1998: Central Administration Legalization of Operational Dosimetrics

Decrees 98-1185 and 98-1186 of 24 December 1998, modifying Decrees 75-306 and 86-1103 "relative to the protection of workers against the dangers of ionizing radiation," are France's official answer to the European Commission's demand on integrating the 1990 directive on "protection of outside workers exposed to radiation during operations in controlled zones."

The definitions of operational dosimetrics given in these texts do not, however, make any less ambiguous the aim of measuring individual dosimetrics, although the decree specifies that the director of a power plant (with the collaboration of the employer) must "first assess collective and individual dose levels," then "measure and analyze radiation doses received during the operation." Does such a procedure lead to limiting exposure, or to stricter oversight of the way dose is shared among workers? The following articles on centralization of information by the OPRI suggest that an implicit choice was made in favor of the second hypothesis. This is confirmed by paragraph II of Article 20*bis* of the decree for having these provisions implemented by persons deemed "skilled in radio-protection" after certification by the OPRI. The only precise indication on application of that skill is having "access to name-specific results of workers' individual exposure levels as measured over a period not exceeding the preceding twelve months" (Art. 20bis II). The fact is that a single "operation," even extended to cover the total length of a unit shutdown, does not usually exceed three weeks. The responsibility of the designated radiation protection inspectors therefore does not include checking exposure at the source, in compliance with the ALARA prevention principle, but instead keeping track of individual name-specific, dosimetric levels in accordance with the "radiation credit" represented by quarterly or yearly dose limits. Indeed, the decree specifies: "as soon as they (the radiation protection inspectors) learn that a worker has reached the limits of legal exposure, they must immediately inform the company doctor or doctors in charge of that worker" (Art. 20bis II). But there is nothing in the decree concerning any guarantee of employment and remuneration for "outside" workers who have reached those limits.

The CHS-CT has access only to non-individual results, a fact that represents a hurdle for this organization representative of in-house as well as subcontracted company employees and which supposedly has the right to know how health in the workplace is being supervised and how owner-operator company and outside employers manage name-specific dosimetric results.

Lastly, the text says nothing about the only clauses that might *structurally* impede managing employment by dose, i.e., adoption of maximum exposure-at-the-source values and maximum average exposure over eight-hour periods (i.e., a normal day at any given work post) such as those operative in other cancer-inducing occupations for which official rules and regulations exist. These are, theoretically, the elements of truly operational dosimetrics. They follow the general principles of prevention, and they should be given priority.

The decree of 24 December 1998 therefore crowns the efforts of nuclear industry leaders to obtain at least administrative legitimacy for this system of selecting workers according to risk levels while freeing themselves from all oversight by labor union representatives.

April 2000: Draft of the March 31, 2003 Decree to Protect Workers Against Ionizing Radiation

This text is the first version of the 2003 decree, which officialized the yearly 20mSV norm for workers. It mentions not only individual dosimetrics but dosimetrics per operation as a means of monitoring exposure. However, the proposed operational dosimetrics are not accompanied by a TLV. The draft indicated that employment must be protected, even that of temporary workers having exceeded maximum dose allowed. The final text of the decree makes no mention of this obligation, however, despite the fact that it is specified in the Code du Travail.[19] Finally, it is important to underline the fact that this is once again only a decree; the nuclear industry still falls outside the realm of legal and parliamentary action. Discussion bears only on equal representation of business and labor on Ministry of Employment and Solidarity advisory committees.

What Sort of Legitimacy?

Through CNIL approval and ministerial decrees, the recording of name-specific data on individuals by means of electronic dosimetrics has gained a certain legal validity. Considering exclusively the problem of not exceeding dose limits, the centralized, interconnected system set up by the nuclear industry and the GIIN is in fact an effective tool for managing distribution of the collective dose of ionizing radiation among a unstable population of "outside" workers while remaining within legal norms: from 1992 to 1998, the number of workers with an individual annual dose over the yearly 20 mSV limit fell from 1200 to 85. The stated aim of the recording system is to manage per-operation dosimetrics in order to lower dose received on the job. However, that recording system as managed by the OPRI did not pertain to on-the-job dosimetric data but cumulative individual doses. This is a deliberate disregard of the files' stated objective.[20]

With respect to labor law and legal protection against abusive use of individual name-specific data, how can we account for the fact that the CNIL approved and deemed legitimate the plan to have a file on employees' performance of their occupation managed by bodies with no precise legal status and offering no specific official guarantee (such as a work contract) to the people involved? The purposes of that file have been misconstrued, and the victims suffering loss of employment and income due to such practices have no legal recourse.

Standing in contradiction to European Union and French legal rules on prevention, health in the workplace and improvement of working conditions,

[19] "Measures concerning safety, hygiene and health in the workplace must under no circumstance entail a financial burden for workers" (Art. L231-11 of the Code du Travail).

[20] To this day, I have received no reply to a letter raising these issues, sent to the president of the CNIL in February 1996.

the rules and practices of radiation protection in France have been placed under the exclusive control of owner-operators and employers in a system of company self-regulation. The contrast between classic administrative "inertia" when it comes to setting up a regulatory system (in compliance with European Union directives), and the "efficiency" of all the protagonists–companies, the GIIN, and the PROMINF–who turned to the CNIL to obtain official approval for managing dose (and thereby employment) of the nuclear industry's outside workers, raises other points of contention concerning the legitimacy of the system thus put in place, not only from a legal standpoint but also from a social and political one.

The extremely sophisticated strategy for monitoring dose received reveals one of owner-operator's major difficulties: figuring out a way of organizing maintenance operations and mastering the problem of exposure to ionizing radiation that would be different from a strategy of dividing up the dose. This situation also calls into question the role of the state in regulation of radiation-exposed work. Through its administrative torpor, the state has given *carte blanche* to the executive managers of a strategic national industry instead of implementing effective protection of exposed workers' health.

Lastly, the choices made in matters of radiation protection are of immediate concern to general safety. As we saw in Part I, maintenance operations during a unit shutdown are now the job of "outside" workers only. The sole preoccupation of the authorities in charge of the nuclear industry is to check compliance with dosimetric norms, regardless of any reference to the substance and conditions of work exposed to radiation. But doesn't that strategy cause the company to lose any means of effectively controlling the sources of high exposure? In the course of maintenance operations, how does the owner-operator compensate for the loss of the professional know-how of workers who have been excluded, some lastingly, from the nuclear industry due to computerized processing of dose data? In managing employment through forced flexibility, how can the owner-operator preserve the concrete memory of the history of maintenance in nuclear power plants?

CONCLUSION

By implementing computerized selection of DATR workers, EDF has successfully circumvented the contradiction between maintenance task performance (which is necessary to preserve safety) and radiation protection. On the one hand, it has acquired exclusive control of social division of the dose without having to worry about managing its effects on the wages, jobs and health of "outside" workers. On the other, it has freed itself from standing constraints in matters of regulation and equal representation on advisory bodies. Lastly, the computerized system has been conceived in such a way as to protect EDF from any likelihood of being taken to civil or criminal court: it is impossible to prove discrimination and, for cases of delayed occupational disease (leukaemia, cancer), the December 1998 decrees do not specify that dose archives must be kept, when of course such archives would enable persons concerned or their heirs to obtain proof of exposure.

The three sources of nuclear industry legitimacy have, to all appearances, been preserved: the industry is competitive and ensures both safety and radiation protection. It remains to be seen, however, what effects this way of organizing labor actually have on health and safety. This is the matter of Part III.

PART III

What Repercussions on Workers Health and Nuclear Safety?

The choices made in the matter of organizing work in the nuclear industry have made it possible to establish a public representation of respectability and social legitimacy for that industry that in turn allows for sidestepping the contradictions inherent in nuclear production. But what repercussions does this social organization of work have on the two public health issues at the core of this industry: the health of workers assigned to perform maintenance tasks, and installation security for the protection of populations around the site?

Social invisibility makes it particularly difficult to identify those consequences. The two chapters of Part III take up the theoretical and methodological questions involved in how knowledge about consequences is developed, and provides components of an answer as to what those consequences really are.

CHAPTER 6

The Health of "Outside" Employees Doing Maintenance Work in Nuclear Power Plant Controlled Zones

> Taking advantage of an individual's vulnerable or dependent position in order to subject him or her to working or living conditions that are incompatible with human dignity is punishable by 2 years of imprisonment and a 5000-franc fine. (Article L225-14 of the French penal code)

> Why not dream, then, of hunting down heterodox genes and launching a Genetic Inquisition? Such dreams, as we all know, are not only dreamt by a few biologists with a philosophical bent. . . . By dreaming such dreams, one enters into another world, on the borders of the brave new world thought up by Aldous Huxley, from which the sick, their particular illnesses and doctors were banned. One imagines the life of a natural population to be like a bingo game in which civil servants are appointed by the sciences of living organisms to verify the conformity of the numbers in the bag, before the players are allowed to pick them out and place them on their cards. At the root of this dream, there is the generous ambition to spare innocent and feeble individuals the terrible weight of having to represent themselves the errors of life. At the finish, one finds the gene police, under cover of Genetic Science. (Canguilhem, 1966, p. 280)

It is no easy matter to identify the effects of work organization on the health of outside company employees performing maintenance operations in nuclear power plants. As we know, ICRP experts have estimated the decrease in life expectancy that may be attributed to exposure to a given collective dose level. By ICRP probability calculations, approximately 1000 man-sievert implied a loss of life expectancy of 880 years for French power plants between 1977 and 1996. We have also seen that 80% of the damage is suffered by outside employees in the nuclear industry. But how can we move from this abstract evaluation, which reifies and thereby masks the reality of death, to actual knowledge of personal rather than anonymous cases, of premature death from been exposed to radiation in the workplace? This was one of the questions our study sought to answer.

There are two types of difficulties in identifying the health effects of the entire set of risks and constraints imposed on outside workers. The effects of low radiation doses are subject to debate. It is important to understand the terms of the debate in relation to the highly specific situation of the population under study. However, our research shows that information on outside workers' health is scattered and difficult to assemble. Further questioning of workers and the company doctors who followed them for a time attests to the social construction of the invisibility of deleterious effects of exposure to ionizing radiation, and of exposure to the risks, conditions and constraints of the way work is organized as identified by both this study and the STED survey (Doniol-Shaw et al., 1995). Speaking directly with "outside" employees is crucial to identifying their health problems, which are not only physical; the issue of psychological suffering also arises in all research exploring relations between health and the different forms of socio-economic instability and precariousness that are increasingly affecting work (Davezies, 1997).

THE DEBATE ON EFFECTS OF LOW DOSES OF IONIZING RADIATION

The controversy about effects of low doses of radiation on health illustrates the value conflict around what is at stake in public health. As work in the philosophy and sociology of science has illustrated, scientific controversies are a way of unveiling the different types of social logic, value systems and representations operative in the social construction of hypotheses and research methods as well as in research objects themselves (Callon & Latour, 1991; Kuhn, 1983).

In public health, two theoretical systems based on implicit conceptions of how health and society are related underlie research seeking to account for epidemiological phenomena (Turshen, 1989). The first system conceives society as the sum of individuals whose particular characteristics—biological and genetic but also psychological, behavioral and social—explain the genesis and distribution of pathologies in a given population. The second system sees society as a historical construction, and health as a process of interaction between the body (which is biological, psychological, but also cultural and social), material conditions (which are also socially determined, by cultural and technical developments related to political choices) and social relations of power and domination.

These two theoretical systems are the basis for different and partly contradictory types of reasoning in the field of public health knowledge and action. In the first, economic and social organization is somehow the "natural" and intangible framework within which the life and health of individuals is necessarily organized. Epidemiological knowledge looks for the causes of illness at the level of individual features, reducing social conditions themselves to mere dimensions—genetic, biological, psychological—of such features among others. The results are then used to determine therapeutic action to be taken on individuals. Those results also lead to selection practices, namely in employment matters. Affected persons become in a way responsible—biologically, genetically but also socially—for their health problems. Michel Foucault and Georges Canguilhem have shown how this representation of public health is at the origin of what Foucault called "biopower" (Foucault, 1976) and what Canguilhem, as early as 1960, identified as the discriminatory logic of a "genetic inquisition" run by "gene police" (Canguilhem, 1991).

The other theoretical system takes into consideration not only individual features but also the dynamic process in which health, material working and living conditions, social relations and collective political choices fit and function together. With regard to action and policy, each component of this process opens up spaces in which social transformation becomes possible, since what is at issue is the improvement or deterioration of people's state of health and reducing or increasing inequality. In this view, public policy means not only informing and training individuals about the risks they take and in individual and collective means of protection from danger, but also developing a critical attitude to existing risks and collective working and living conditions, and acting on that critical understanding.

How do these conceptions determine choice of research approach and policy action in matters of radiation protection?

The strategy adopted for radiation protection is based on the implicit hypothesis that a danger threshold exists which should determine official radiation dose limits. The terms of the controversy surrounding this hypothesis are rooted in the value conflict mentioned above, and in the international history of the nuclear industry. The most recent debates on this question revolve around the health effects of the Chernobyl accident, discussion of the abnormally high number of leukaemia cases among persons under 25 living near factories treating nuclear waste, and the relation between occupational exposure to ionizing radiation and cancer. It is worthwhile giving a quick summary of those debates since they have determined the framework for discussion of effects of the exposure to which the nuclear industry's "outside" employees are subjected.

The Consequences of Chernobyl

Without going into detail on the debates among experts on the extent of damage caused to public health by the Chernobyl accident, it must be stressed that the controversy has not led to precise in-the-field assessment of the facts. No consistent, on-going medical follow-up was organized after the catastrophe for observing how the health of the different affected population categories evolved; i.e., that of the "liquidators" who worked on the highly contaminated site shortly after the disaster;[1] contingents from all over the former Soviet Union who were mobilized to decontaminate the soil and slaughter cattle and farm animals; population groups evacuated from the zones near the power station, now off-limits to residents; inhabitants of more distant zones, not only in the countries of the former Soviet Union but also in all the other countries affected by radioactive fallout from the Chernobyl cloud, including France (Belbeoch & Belbeoch, 1993). No conclusive appraisal was made, and worse yet, none can ever be made given the extreme dispersion not only of the radioactivity itself but also of the exposed populations. No particular follow-up was put in place for workers assigned to operate in and decontaminate the site or its surroundings. Given the vast diversity of tasks performed during the state of emergency, with the barest protection and no precise determination of radioactive elements present and dose levels, "estimates" of yearly average dose say absolutely nothing about the reality of actual exposure. According to the OECD's Nuclear Energy Agency, the number of "liquidators" "remains to be precisely ascertained; as of this date, only a few data have been collected." The suggested figure is 800 000 (NEA 1995). Eye-witness accounts of events that occurred over ten years earlier are presented by the Belarusan writer Svetlana Alexievitch in *Voices from Chernobyl: The oral history of a nuclear disaster* (2005). These show that there was no clinical supervision or follow-up of the people irradiated in Chernobyl, particularly the most severely exposed; i.e., the workers themselves. The upheaval in the lives of all inhabitants of the region, whether they remained in the contaminated zones or were forced to leave, represents an insuperable obstacle to organizing any after-the-fact follow-up, especially given the political, social and economic changes that have taken place in the former USSR. Any attempt to validate the hypothesis of a correlation between morbidity and Chernobyl-related occupational and environmental exposure will therefore always run up against poor post-Chernobyl health surveillance and lack of systematic or rigorous data.

Abnormally High Numbers of Leukemia Cases among Persons Under 25

Since the early 1980s epidemiologists have been debating the reasons for the unusually high number of leukemia cases among children living near the Sellafield and Dounreay nuclear reprocessing plants in Great Britain. The terms of the debate clearly reflect the influence of the two conceptions mentioned above of what research in public health should be. The debate does not bear on the number of cases itself, which nobody denies, but the risk factors involved. In the following interpretation, put forward against the hypothesis of a causal link between the abnormally high number of cases and exposure to ionizing radiation (Gardner et al.,

[1]Some liquidators went on a hunger strike in 1990 to force recognition of the radiation-related health problems they were suffering from (Belbeoch & Belbeoch, 1993).

1990; Urquhart, 1991), it is stated that the high number of leukemia cases among persons under 25 living near the English factory is due to "an increased risk of viral infection due to the mix of populations of diverse origins brought together in building the power plants" (Kinlen, quoted by Doll and his research group in Forman, 1987 and by Hill, 1995). The first study of radiation effects among the population living in the Nord-Cotentin region of France (Normandy) shows a number of leukemia cases among persons under 25 for the 1978-1993 period that is significantly abnormally high; the geographic distribution of these persons points to a concentration in counties near the La Hague nuclear reprocessing plant (Viel, 1993). Publication of the results of Viel's case-control study in 1997 reawakened debate. Starting out with Gardner et al. (1990) and Urquhart's (1993) hypotheses and method, this study shows a statistically significant relation between the appearance of leukemia and non-Hodgkinian lymphoma and the following risk factors: being frequently on the beach and consuming shellfish. It concluded that further epidemiological research was needed into the hypothesis that radioactive environmental pollution due to waste from the La Hague nuclear reprocessing plant might affect the likelihood of leukemia developing among children in exposed populations.

This study triggered very sharp and immediately negative reactions against Viel from other researchers, duly relayed by the newspapers, though no scientific, critical discussion of the study that might have invalidating its results was forthcoming (Viel, 1998). At the behest of the public administration, a study on the incidence of leukemia was conducted in the same region from 1993 to 1996 (Guizard, Spira, Troussard, & Collignon, 1998). Although 8 new cases were discovered, the authors concluded that the number was not significantly abnormally high, and proposed a new hypothesis to explain Viel's resuts: they were due not to radioactive environmental pollution as such but rather to this kind of pollution as one of many individual characteristics of the population observed: "The results pertain to the entire population of the ten counties and are not incompatible with the fact that the number of leukemia cases appears to have increased among persons who share certain characteristics (frequently visiting Nord-Cotentin beaches and consumption of locally caught fish and shellfish). It is possible that those characteristics only attest to the diversity of types of exposure in the population: ways of life, exposure to ionizing radiation of natural, artificial and medical kinds, to chemicals, to viral infections, etc."

When presenting the papers of the *"Nucléaire et Santé"* (Nuclear industry and health) conference on January 30, 1998—it was at this conference that the counter-study was presented—Professor Tubiana broadened the scope of the counter-study's results: "Contrary to what was announced in the press, there has been no increase in the amount of leukemia around La Hague. That's a fundamental fact." However, he neglected to indicate that the 1993-1996 results referred to studies done at a later time and using a different method from the previous ones. Finally, in the opinion of its own authors, the results of the later study are not incompatible with the abnormally high number of cases brought to light by Viel's study. However, in the same publication, the study results of the English researchers together with Viel's are disqualified with the phrase "*Huge epidemio-media wave at La Hague.*" After this episode of disqualification, Viel's results were never integrated into scientific debate on the effects of low radiation doses.

According to Alain Desrosières, an eminent specialist in the field of statistics,[2] all statistical work is a "black box" that must be questioned with the idea that "if we want to reflect on the nature of statistics, we must never separate them from the use they are being put to." In order for probability of a relation between a type of cancer and exposure to a risk factor to be perceived, the statistician's "black box" has to contain a certain representation of the reality under observation, a representation which will then serve as a guide for formulating hypotheses and choosing variables. That representation is socially constructed with reference to postulates that found a theoretical system for interpreting facts. The same factor may be associated with different effects depending on the system of interpretation that factor is embedded in (Bourdieu, 1979).

In the controversy opposing Spira and Viel, Spira subscribed to the first system for interpretating illness, oriented toward looking for the cause at the individual level, whereas Viel was critically concerned about collective risk factors as indicators about the surrounding area and changes in it.

What is missing in this controversy is fundamental information that would allow for comparing morbidity facts to facts about risk and danger—that is, precise knowledge about exposure to radioactivity. This would mean knowing the exact amount of radioactive elements present on the beaches children go to and in the shellfish they eat, as well as how this evolves over time. The same is true here as what was underlined for DATR workers: health authorities lack concrete measurements of occupational exposure to ionizing radiation; there is no systematic, verified data on exposure itself that would allow for determining the nature of radioactive elements and level of radiation. In its memo on methodology with regard to the impact of La Hague's nuclear-waste reprocessing activity, the Groupe Radioecologie of Nord-Cotentin, presided by Sugier, complained that the extremely limited number of measurements did not allow for any detailed assessment of environmental impact (*Groupe radioécologie Nord Cotentin,* 1998).[3] Comparing the organization of public health in France with that in other countries of Northern Europe, Viel has pointed out that in France

[2]Interview in *Le Monde,* September 29, 1998.

[3]The taskforce had therefore based its assessment on theoretical models and concluded it was theoretically impossible for there to be an inordinately high number of leukemia cases associated with exposure as extrapolated by the models (*Le Monde,* July 9, 1999). But how closely do these models reproduce the reality of the nuclear waste released into the environment by the La Hague plant?

there are no arrangements that allow for any linking of possible exposure to radiation and health damage: exposure levels are not measured and health damage is not recorded with an eye to future developments (Viel, 1998).

Spira, on the other hand, subscribed to an etiological research perspective in which hypotheses focus on biological, genetic and behavioral causalities and risk depends first and foremost on individuals themselves. As we saw, the report's proposed research program for Nord-Cotentin aims to explore the hypothesis of particular genetic vulnerability or the existence of a virus transported in the course of group movements at the time the plant was being built. This implies arrangements for surveying the population and presupposes access to files that up until now have been inaccessible because protected by the law of Sécurité Sociale ID numbers.[4]

The proposals contained in Spira's report on steps to be taken by epidemiologists in relation to occupational health for workers exposed to ionizing radiation primarily concern statutory, permanent nuclear industry workers (COGEMA, CEA, EDF) for whom medical follow-up already exists. But no morbidity study is proposed for the population subjected to 80% of exposure. In the International Agency for Research on Cancer (IARC) study of mortality amongst nuclear power plant workers, all subcontracted workers should have been integrated into the survey (Cardis et al., 2005). Methodological choices have been made that are not compatible with precise statistical identification of correlations between deaths and particular working conditions of employees. Workers from companies with fewer than 50 employees were omitted at the outset, although they are often the ones assigned to the most highly exposed work posts over short lengths of time because they come at the end of the subcontracting chain. Exposure is indicated in terms of an average yearly value, which has the effect of blurring all the specificities of the radiation actually received. Internal contamination through the inhalation or ingestion of radioactive elements is not taken into account. The time lag is insufficient vis-à-vis latency periods for cancer among a young population whose highly diverse occupational itineraries, tasks, and types of exposure make it particularly difficult to explore statistical correlations concerning risk of ionizing radiation, "all else kept equal."

Meanwhile, using general cancer registers to identify abnormally high numbers of cases that might be connected to ionizing radiation is not an appropriate means for identifying occupational cancer, since the records of those registers contain no data on patient's occupational history and they in no way allow for distinguishing occupationally exposed groups from groups who have been exposed to radioactive environmental pollution and groups who have not been exposed. The fact that DATR workers are scattered throughout France also

[4]The CNIL granted Spira permission to consult these files, something that no other researcher before him had even obtained.

disqualifies any geographic method that would take into account "distance from nuclear sites" only without considering nature and level of exposure.

Ethical and political questions are also raised by scientific choices, questions about the repercussions of research results: What benefits can be expected, *for affected individuals,* from research proposed in the Nord-Cotentin on the genetic or viral etiology of observed leukaemia cases? How will such research respond to the fears of people exposed to ionizing radiation, and what will its answers be? Canguilhem was already demonstrating in the early 1960s that the effect of research into genetic vulnerability would be to stimulate strategies for selecting (stigmatizing) persons who carry genes that expose them to certain risks. Will prevention consist in testing people and then forcing genetically radio-sensitive persons to move away from Nord-Cotentin? After such research is done, will genetic selection replace managing employment by radiation dose in the way radiation-exposed work is organized? Is genetic "marking" on the way to becoming a motive for disqualifying individuals professionally and socially, without any public debate on the issue? Wouldn't this also be the best way to let nuclear industrialists off the hook, freeing them not only from all accountability for safety and protection against radiation, but also from having to compensate victims? The International Nuclear Energy Agency's own Committee on Radiation Protection and Public Health has raised just these questions about the uses to which research results on genetic predispositions might be put:

> Particularly for workers, the question of being genetically predisposed is likely to become one of the important subjects of radiation protection as it may have an impact on employment policies and employability, as well as insurance and compensation. However, if science proves one day able to provide useful information, the use to which it will be put must be decided beyond the small world of radiation protection; it is the responsibility of other segments of society. (Committee on Radiation Protection and Public Health, Nuclear Energy Agency, 1998, p. 26)

Occupational Exposure and Risk of Cancer

Since the early 1980s, particularly in the United States, Great Britain, and Canada, an increasing number of studies of nuclear industry workers have suggested that assessments based on the results of cohort studies of Hiroshima and Nagasaki survivors—the reference model for the ICRP—may have substantially underestimated the risk of cancer linked to low levels of occupational radiation (Cardis et al., 1995; Wing et al., 1993; Wing et al., 1999). However, those results have not been used to lower recommended levels of exposure to ionizing radiation. The controversy on effects of small doses has flared again, but for a different reason than the ones mentioned. Comparing occupational epidemiology studies with those on atom bomb survivors in Japan, Wing et al. (1999) bring to light the systematic differences between study populations, sample sizes, types of exposure to radiation and measuring methods—study results and factors likely to generate confusion. The fact that the observed differences were recurrent led the authors to raise the issue of the relevance of the epidemiological model used to study Japanese survivors to account for relations between radiation and cancer in the case of nuclear industry workers. Questioning the epistemological reasons for choosing that internationally accepted model, they retrace the history of epidemiology of occupational cancer in the American nuclear industry, showing that it developed extremely late in relation to expansion of both military and civil nuclear industries in that country and the start of survivor cohort studies in Japan. Absence of effects of "such small doses" as those received by workers in nuclear installations was posited and accepted, without any prospective study on the health effects of continuously received radiation. The model resulting by extrapolation from the Hiroshima and Nagasaki results became dominant, instilling doubt about any results not in accordance with it, including those for cohort studies conducted since then in many other countries.

In France, epidemiological study of mortality amoung uranium miners in the Limousin region evidenced an abnormally high rate of cancer.[5] At the present time, the socially constructed invisibility of occupational illnesses related to nuclear industry work can be explained by the way employment and radiation dose are managed in the nuclear industry, both in France and other nuclear countries, particularly Japan (Tanaka, 1997).

PROTRACTED FOLLOWING OF SUBCONTRACTED DATR EMPLOYEES IN THE NUCLEAR INDUSTRY

Undertaking a study of the effects of working conditions and exposure to ionizing radiation on workers' health presupposes identifying and finding all exposed workers and being able to follow them over time. After conducting a survey in 1988 of 87 DATR workers concerning how they were medically and administratively followed, and in order to test the feasibility of such study, we reestablished contact in 1994 both with the workers and company doctors of the previous survey. The aim of the second survey was to situate the results in a longitudinal perspective that would take into account how these workers had fared in regard to health and employment. No such study had ever been done. The complementary survey was also aimed at evaluating whether it was possible to institute real and effective medical follow up.

Only six years separated the two surveys, and the workers had been informed that the first one was likely to have a sequel. Yet we could not contact any more

[5]These results, presented to a taskforce of the Office Parlementaire d'Evaluation des Choix technologiques [Parliamentary committee for the evaluation of technological choices] in 1999, have never been published.

than half of the original group. The company doctors had lost sight of the other half of the workers, and the information they have on health and employment is very fragmentary. However, on the basis of research results from other studies (Davezies, 1995; Dessors & Volkoff, 1992; Frigul, 1997), we can hypothesize that it is not by chance that certain employees are still in touch with the medical services and others are no longer in touch. The observed instability and "precariousness" of workers' situations also makes the matter of collecting and updating information on their health a precarious matter.

What, then, are the real, practical possibilities of implementing occupational and post-occupational follow-up of workers who have been exposed during their occupational activity to one or several risks of cancer, keeping in mind that medical follow-up of this kind is required by the regulations.

What is meant by occupational and post-occupational follow-up of DATR workers? From the perspective of the owner-operator EDF, we know that the only thing that matters is dosimetric follow-up, to prevent a worker from exceeding the maximum dose. What workers wish and expect from medical follow-up is of an entirely different order. For them, follow-up should be medical accompaniment that allows for the earliest possible detection and, if necessary, treatment of pathologies connected to radiation or contamination—leukemia and cancer, but also sterility and various other afflictions of the eyes, thyroid gland, lymphatic or digestive systems, for example. For workers who have been exposed to a risk of cancer, it should also mean being able to access information and obtain the necessary medical certificates for obtaining free medical follow-up and, in case of leukemia or cancer, official certification of occupational illness. Lastly, it should mean psychological as well as clinical, medical, and legal assistance. Psychological help is critical to calming anxiety[6] and preventing situations that threaten workers' continued employment. The fact is that being declared medically unfit due to damaged health raises the threat of being declared permanently occupationally incapacitated. For pathologies associated with ionizing radiation that require the decision of a company doctor, if this doctor is unaware of the worker's "radioactive past" (and thus of the connection between occupational activity and the pathology), the worker's unfitness will not be "compensated" with certification of an occupational disease; he will not be able to obtain precisely the declaration required to be eligible for compensation. One such reparation right entails protecting the employment of persons who have become victims of a workplace accident or an occupational illness.

With respect to these various aspects of medical follow-up, the information collected from company doctors—referred to below as "institutional

[6]A result of the statistical survey of outside workers in the nuclear industry that CRG researchers at Ecole Polytechnique found particularly significant was the expression of anxiety as a response to nuclear danger (Dégot et al., 1999).

knowledge"—essentially amounts to overseeing dose received regardless of the specific working and exposure conditions experienced by workers. But direct interviews with workers has revealed that serious health problems connected to their jobs appeared during the six years separating the two surveys. Company physicians either never knew of those problems or chose to ignore them.

Occupational Physicians' Institutional Knowledge[7]

EDF's company doctors are in a paradoxical position. The Decree of February 1992 on outside workers operating in plants makes them responsible for the special medical surveillance of workers exposed to specific risks generated by plant conditions themselves. This should apply to employees working in nuclear power plants, i.e. where the risk of exposure to ionizing radiation is generated. But as explained, a special dispensation has transferred responsibility for workers exposed to ionizing radiation to the company doctors of the outside firms employing those workers.[8] EDF doctors therefore do not have information concerning outside workers' health. The only information to be had from them in the framework of our survey was the number of DATR workers in the first group who were still, in 1994, in the DOSINAT file. In January 1994, this was the case for 59 of them, or 68.8% of the original survey population of 87.

There are many *inter-company doctors and doctors working for independent subcontracting firm services* who are in charge of medically following these workers, and they themselves are widely dispersed. It is useful to recall the results of the 1988 survey on medical follow-up of DATR workers by outside company doctors:

[7]Occupational medicine in France is a private institution. The Code du Travail stipulates that every company must use occupational medicine services. The biggest companies, such as EDF, have their own company doctors and services. Small and medium-sized businesses (an official category) contract with inter-company institutions for the services of occupational doctors. Each occupational doctor at EDF follows approximately 500 in-house permanent employees. Inter-company occupational doctors , however, follow an average of 3500 employees working for 200 to 300 companies.

[8]The fact that subcontracted companies are now required to obtain nuclear certification in order to do medical follow-up on DATR workers has not fundamentally changed how such follow-up is done nor has it altered the modes of subcontracting and recourse to temporary labor which are at the root of the extreme dispersion of information on workers' health. In the event of an outside company's not acquiring certification, Decree 97-137 of February 13, 1997, stipulates that EDF company doctors must provide special medical supervision for DATRs. In any case, the fact that workers frequently have a different doctor due to different jobs and/or changes in company makes on-going follow-up extremely difficult.

DATR Survey
(Thébaud-Mony et al., 1991)

- 103 different doctors participated in compulsory medical surveillance of the 87 DATR employees over the 4-year observation period. In 40% of the cases, a worker had one doctor; in 17%, two doctors; and in 43%, the number of doctors per employee ranged from 3 to 6.
- The average length of follow-up by the same doctor for employees in the study was less than two years.
- The medical itineraries of outside employees fall into three categories:
  - those whose medical surveillance has been handled exclusively by an independent or inter-company medical service of which the company is a member (41 employees, or 47%);
  - those whose medical surveillance has been handled by both the service with which the company has a contract and other medical services (27 employees, or 31%);
  - those whose medical surveillance has been handled exclusively by medical services with which the company has not signed any agreement (19 employees, or 22%).

Such differences in the organization of mandatory medical surveillance suggest that it is virtually impossible to speak of "medical follow-up" in the situation observed.

The extremely dispersed character of medical surveillance in the case of these workers meant that many of them had been "lost sight of" by 1994; information was forthcoming from outside company doctors for only 43 out of the 87 cases (below 50%). Among employees still known to the industrial medicine services, 34 (that is, 40% of the total in the original study), were still being followed by the same service, though not always by the same doctor. In 9 cases (less than 10%), the company doctor knew that the employee had changed companies, and for some the file had even been forwarded. Clearly these doctors' information was extremely limited: none of the company doctors interviewed mentioned the notion of serious illness or death.[9]

[9]This part of the survey was done *via* labor physician-inspectors (who are contracted by the central labor administration as advisors for labor inspectors specialized in health issues) and was thus covered by the privacy clause for medical professionals officially charged with protecting health in the workplace.

Acquiring Knowledge of Employees' Present State through Direct Contact With Them

The occupational itineraries of the employees participating in the study were typified by employment instability, "precariousness," and mobility. This had already been the case in 1988.

Occupational Itineraries and Employment Flexibility in 1988
(Thébaud-Mony, 1991)

- Each type of activity described for maintenance in the nuclear industry is represented.
- Fewer than 50% of survey respondents were employed by a single firm, either permanently or on an intermittent basis. Of the rest, 17% were new DATR permanent employees (in 1988), 14% worked for temporary agencies, and 20% had worked consecutively for several companies, either on a permanent or intermittent basis.
- Total number of companies mentioned was 68, 6 of which are temporary agencies. 38 of the 68 companies were mentioned by only one employee.
- Occupational itineraries differ by individual worker and type of activity.
- Average number of moves between different power plants is 3 per employee per year (maximum 21). Frequency of changes goes together with long distances between power plants, and between plants and place of residence. Altogether, the survey population resided in 28 different regions of France.

Renewing the survey through direct interviews of workers themselves or their families allowed for gathering more precise information on later employment and health of those who could be tracked down. At the end of the first survey, a letter had been sent to DATR respondents informing them of the survey's main results and that it might be continued—as indeed it was in 1994. This time respondents were interviewed by means of a questionnaire, either by telephone or by mail. This direct contact, sometimes with respondents' families, brought to light the situations of 35 workers.

Ultimate Information on the Situations of the 35 Workers Whose Later Itineraries were Described to Us By Themselves or Their Families

- 5 had died (circumstances of death to be presented further on);
- 18 had remained DATRs; 8 of these were still working for the company they were employed by in 1988; 10 had changed employers; 3 were working as intermittent DATR employees in 1988 and still doing so in 1994;
- 4 were still employed by the company they were working for in 1988 but no longer assigned to radiation-exposed work;
- 6 had definitively left this sector of activity and were working in other sectors;
- 2 were unemployed.

Altogether, 12 of the 35 workers were still at the company they had been working for in 1988. It is reasonable to hypothesize that this constitutes the only fraction of truly stable employment among the population surveyed in 1988 (i.e., 13.7%).

The above results confirm the intense employment flexibility to which DATR employees are subjected. Whatever the nature of their work contract, few benefited from stable employment in the same company from 1988 to 1994, even when continuing to work in a nuclear power plant and exposed to radiation.

It must be stressed that, of the 12 temporary workers present in the 1988 survey, 8 could not be found, 3 were still working on a temporary basis and directly assigned to radiation-exposed work, and one had died. The 4 temp workers found in 1994 (including one who died) were confined to the instability and precariousness of that particular form of employment. This "new form of employment" (according to current vocabulary) would more fittingly be called "substitution of temporary for permanent employment"—and such substitution is illegal according to the Code du Travail. These are now "loyal" temporary workers [meaning they work in nuclear power plants only, but on temporary contracts], performing their tasks in one or several companies and doing the dangerous jobs. It is true that seasonalizing unit shutdowns allows for interpreting the rules in a way that justifies the existence of "activity peaks" that require resorting to temporary employees. In light of the system of management by radiation dose presented in the previous chapter, that strategy appears above all like a means of circumventing radiation protection rules, since at any moment temp workers can be thought of as having "finished" their "job." Moreover, hiring through a temp agency is the mode of temporary employment preferred by subcontracting employers in the nuclear industry, because, as they say, it is "the most flexible" for meeting owner-organizer company requirements.

Comparing Information from the Two Sources: Institutions and the Workers Themselves

To evaluate the real possibilities of medically following this population, I compared the data gathered from three sources of information.

Institutional Medical Knowledge

- 23 employees (26% of the original sample) for whom a company doctor had provided information answered the questionnaire either by telephone or in writing;
- 20 employees (23% of the original sample) for whom a company doctor had provided information did not answer the questionnaire;
- 12 employees (14% of the original sample) were unknown to company doctors but the questionnaire was answered, either by the employee himself or his family. Among them, 5 had died; one had had two workplace accidents; and two were out of work;
- 32 employees (37% of the original sample) never answered the questionnaire and company doctors had no information on them.

Outside company doctors do not have direct access to DOSINAT data. Due to the great number of companies and nuclear sites and their geographic dispersion, there are many doctors spread out all over France. A discrepancy exists between number of cases repertoried in DOSINAT by EDF company doctors (59) and number found via direct questioning of doctors in charge of following those employees (41). The fact that there are more in the DOSINAT file does not tell us anything, however, about either the modes of these workers' temporary employment, the conditions they worked under, or their possible health problems.

Direct questioning of workers or their families shows that the most critical health situations escape the attention of the company doctors whose job is to ensure continuity of medical surveillance and thus make it possible to set up post-occupational follow-up. Furthermore, a large proportion of employees from the survey have been totally lost track of.

The gap between the extremely fragmentary institutional knowledge and the amplitude of collective damage expected from the fact of ionizing radiation alone gives us an insight into how the social invisibility of effects of working under exposure to radiation is constructed. In the structural organization of maintenance work in nuclear power plants, the rules for occupational and post-occupational medical follow-up that figure in the Code du Travail are not complied with—and it would seem they cannot be. Employment flexibility, combined with institutional fragmentation of medical surveillance, causes a considerable proportion of the workers in question to disappear from the field of occupational and post-occupational surveillance. Some of the selection mechanisms, which cause these "disappearances," can be observed by directly questioning employees. However, given their extreme mobility and the fact that some have no stable place of residence, many could not be found. In the absence of regular surveillance, the damage inflicted on their health escapes the attention of those whose responsibility it is to prevent such damage and declare their situation so they may obtain financial compensation.

STRUCTURAL LOGIC LINKING HEALTH DAMAGE TO WORK ORGANIZATION; PREMATURE DEATHS

As mentioned, five deaths figured in the information collected during the second phase of the survey. The families gave information as to the cause of death.[10] The issues raised so radically by these deaths were in fact also raised in the STED survey (Doniol-Shaw et al., 1995, 1998), whose results bring to light

[10] A second questionnaire bearing on the circumstances of death was sent by the labor physician-inspector to the company doctors in charge of these employees' medical surveillance in 1988. The questionnaire was filled out by two company doctors. An interview was done with families.

the structural character of the social differentiation of risks and constraints borne by "outside" DATR employees, and the possible impact these risks and constraints may have in deaths, according to three broad categories of damage to health in the workplace defined by Davezies (1998): "direct damage to physical integrity" due to physical and chemical substances; "damage to men and women made to work inappropriately or excessively"; and "damage to individual worker's dignity and self-esteem." The history of each of the deceased workers is both unique and reflects a shared social condition, the structural measure of which is made clear by the STED Survey, a statistical survey of several thousand outside DATR employees.

The STED Survey

The STED ("*Sous-traitance* [subcontracting], *EDF, DATR*" (Doniol-Shaw et al., 1995, 1998, 1999, 2003) bears on the working, living and health conditions of DATR employees. (Certain STED results have been cited in previous chapters.)

Survey Method

This survey was coordinated by 10 regional-level physician-inspectors and conducted by 140 volunteer company doctors, after a preliminary survey of 850 company doctors working for subcontracting companies or temporary agencies; doctors called upon in recent years to carry out medical check-ups as part of follow-up of DATR workers. The survey was conducted between May and November 1993 in the industrial medicine services responsible for supervising "outside" DATR employees. The sample was chosen by taking into account 1) number of DATR employees likely to figure in the group of workers each doctor followed, and 2) distribution of surveyed workers by specialization in the nuclear industry. Employees themselves filled out the survey questionnaire, accompanied by a card filled out by the company doctor during the medical check-up. In 1998, a second phase of the survey was conducted: 40% of the sample had been lost track of, and a greater proportion of employees had unstable rather than permanent-contract employment (Doniol-Shaw et al., 2001).

There are two sorts of possible biases in the survey. On one hand, a selection effect with regard to both doctors and employees surveyed, due to the fact that volunteer company doctors determined the sample. Specifically, very few company doctors who were following *few* DATR employees—especially those following workers from temporary employment agencies—participated in the study. Our own survey showed that a relatively high number of company doctors were in charge of only one or two DATR employees and even then sometimes only sporadically. That bias would seem to indicate that the number of employees in the most precarious employment situation was underestimated. The other type of bias is linked to time and place the questionnaire was conducted, i.e., in the industrial medicine offices during DATR employees' biannual medical fitness tests. This may have induced self-censorship on the part of employees concerned about the impact their answers might have on the medical fitness decision. The responses to questions about work satisfaction, work experience, and health may also have been subject to that bias, though it is not possible to evaluate such effects precisely. As in the working-conditions survey that accompanied the French Ministry of Labor's employment survey, questions concerning working conditions and constraints are based on perception of risks and the experience of constraints, perception which is in turn dependent on different influences, whether these concern workers' knowledge of the nature and effects of risks incurred or the reality itself and the "objective" characteristics of those risks and working conditions. Gollac (1997) has shown that these biases cannot account for all results of the latest national surveys on working conditions. The same applies to the STED survey, whose results with regard to risks, characteristics of labor organization, and working conditions are of great significance, providing an extremely clear view of the conditions and constraints which are the lot of DATR employees.

STED statistics allow for demonstrating the structural nature of the social division of labor and risk, and of working conditions and health in the workplace. They also elucidate the logic prevailing in the instances of death identified by our research.

Social Division of Labor and Radiation

The statistical study of dosimetric data brings to light that the average yearly dose of radiation is socially differentiated by specialization, type of employment and time spent working at the nuclear site. The highest doses are received by service company employees working in pipe lagging/montage, nuclear servitude, and valves and pipes; i.e. by those who prepare the ground for more skilled workers, enabling them to operate in decontaminated spaces (cf. Part I). Temporary workers (particularly those working for temporary agencies and workers with contingent contracts, present on the nuclear site for the shortest length of time) receive a significantly higher average yearly dose than employees with permanent work contracts.

Yearly average dose does not suffice to tell us precisely who the persons exposed to ionizing radiation are and what typifies their exposure or, barring that, the nature of the doses they receive, and the figure even appears to cause considerable statistical bias. STED survey researchers concentrated not only on establishing average exposure level but also various types of measurements (average yearly and monthly doses, maximum yearly and monthly doses, determination of dosimetric categories). Their analysis reveals extreme diversity in dose "profiles," a diversity that nevertheless follows certain specificities by type of maintenance activity and type of employment. The first type raises questions about the exposure itself, while the second tends to confirm the qualitative results presented here on how employment is managed by dose.

In the STED survey, the strategy implemented by EDF of dividing up dose received does indeed appear closely linked to the social division of labor brought to light in workers' accounts. Yearly dosimetric levels of workers performing service activities (lagging-montage-scaffolding; nuclear servitude; some valve work) are significantly higher than those of employees performing more skilled work. Persons receiving significantly higher average yearly doses are workers as opposed to technicians and managers and temporary workers as opposed to permanent-contract ones. Workers with the highest average yearly doses are temporary workers, temporary agency workers, and contingent contract workers. Temp workers also spend the least amount of time on nuclear sites (less than 6 months a year).

Distribution of workers by dose typology, ranging from absence of exposure to over the stipulated maximum of 20 mSv/year, points up two other important results. First, a considerable proportion of employees undergo very little exposure or none at all (36.1% of the workers interviewed had 0 dose for the year preceding the survey and 31.4% less than 5 mSv). Second, the distribution of DATR workers by dose category points up socially differentiated radiation:

- 11.7% of workers receive average yearly doses exceeding 15 mSv.
- 8.1% of contingent-contract employees receive average yearly doses exceeding 20 mSv.
- 11.8% of laggers, 8.3% of the valve specialists, and 7.7% of the nuclear servitude employees also receive average yearly doses exceeding 20 mSv.

The statistically significant social differentiation of radiation disproves the hypothesis that management of radiation received is explained by the individual responsibility of each employee or company. On the contrary, it supports the hypothesis that distribution of jobs involving the highest radiation doses is not random. Nevertheless, there is no data, either in the STED survey (the researchers did not have access to job assignment information) or in the EDF or OPRI documents (which are confidential), that allows for checking the hypothesis of variation due to type of work performed and exposure itself.

The study also showed that average monthly dose is significantly differentiated by type of employment: temporary workers receive the most radioactivity in the shortest spans of time on the nuclear site (in some cases extremely short: 30 mSv in 15 days). These data, which correspond well to certain employee accounts in Part I, confirm that the strategy chosen by the owner-operator EDF is to divide up the dose among workers. This is only possible if linked to job flexibility and by reducing time worked on the site. This in turn allows for reducing individual yearly average dose for the jobs most exposed to ionizing radiation. The owner-operator thus oversees compliance with the norms but has no idea and pays no attention to how the subcontracting firms themselves are managing employment, working conditions, and time constraints for employees who are close to the official dose limit, since these workers are not under contract with EDF. The results of the STED survey on this point and its effects on outside workers' health also echo the analyses developed in Part I.

Damage Linked to Exposure to Physical Risks and Chemicals

Radiation and other constraints such as carrying heavy loads, noise, exposure to toxic products and particles, prolonged or repeated uncomfortable positions, repetitive movements or operations, working in extreme heat–the frequency of such conditions was brought to light by the STED survey–combine and constitute so many direct attacks on the physical health of "outside" DATR workers. The different types of constraints are significantly more frequent in service activities (nuclear servitude, valve work, lagging).

Immediate, specific effects are not instantaneously observable, particularly where ionizing radiation is concerned, or toxic products that induce illnesses—most notably cancer—over the long term. Nonetheless, the STED survey shows correlations between certain constraints (carrying heavy loads, exposure to particles and toxins, noise) and the manifestation of digestive disorders, respiratory problems and back trouble. Davezies (1998) stresses that these problems "develop into more insidious forms, either because they concern long-term effects or because the damage is not very specific, so the connection with work performed is harder to prove than for traditional types of health impairment." Two of the deaths of "outside" workers in my survey, deaths due to cancer, fit into this category:

A valve worker aged 39 died from bone cancer in 1992. His wife transmitted the relevant file to us, which contains a posthumous request for official acknowledgment of his illness as an occupational disease. A certificate recognizing occupational disease and incriminating ionizing radiation was established by the victim's family physician:

> I, the undersigned Dr M. do hereby certify that I treated Monsieur S. for an illness that might be classified among Table 6 occupational diseases.[11]
>
> The patient was admitted to my hospital ward in November 1991 because of severe cervicalgy [neck pain] and impaired motor capacity in the upper limbs. Examination revealed malignant osteolytic lesions at the level of lower vertebrae and upper spine. A local intervention showed probable metastases of adenocarcinoma; despite extensive exploration, the initial tumor could not be found.
>
> Given this rapidly developing affection in a young patient and the notion of exposure to ionizing radiation, it is reasonable and legitimate to look for a causal relation between exposure to ionizing radiation and this cancerous disease. Classifying this illness among Table 6 occupational diseases seems all the more rational and warranted given that the original lesion could not be

[11] In France, workers can be compensated for diseases linked to working conditions if those diseases are on the official list of occupational diseases for which compensation is authorized or if it can be proved that their disease is linked to their work. Table 6 lists occupational diseases related to radiation exposure (see Réseau Scop93, 2005; Thébaud-Mony, 1994; Table 1in this chapter).

found: the medical literature mentions a great number of different neoplasic affections caused by ionizing radiation and the degenerative nature of the disease; in this case leading rapidly to the death of a subject who seemed previously in good health.

The compensation board at first rejected the declaration filed by Monsieur S.'s widow, stating the following reason: "The illness described is not in the table of occupational diseases." Mme S. made a second request, this time based on the resemblance between her husband's case and that of a decontaminator at La Hague who had been found to have cancer in the soft tissues (muscles) of the shoulder (not included in Table 6 either), whose illness had been recognized by another compensation board, which had taken into account both exposure to ionizing radiation and the data contained in the medical literature. This appeal was also turned down, on the same grounds. Mme S. did not pursue the procedure.

Table 6. List of Compensable Occupational Diseases
Afflictions Caused by Ionizing Radiation

Diseases and time-lapse between the exposure and the disease during which the disease may be recognized as an occupational disease related to ionizing radiation	List indicative of jobs that may provoke these diseases
Anemia, leucopenia, thrombopenia, or hemorrhage syndrome following chronic radiation; 30 days Blepharitis or conjunctivitis; 7 days Keratitis; 1 year Cataracts; 10 years Acute radiodermatitis; 60 days Acute radioepithelitis of mucous membranes; 60 days Chronic radiolesions of mucous membranes; 5 years Bone radionecrosis; 30 years Leukemia; 30 years Bronchial-lung cancer by inhalation; 30 years Bone sarcoma; 50 years	Any job involving exposure to X-rays or natural or artificial radioactive substances or any other source of corpuscular emission, namely Extracting and treating radioactive minerals; preparing radioactive substances; preparing radioactive chemical and pharmaceutical products; preparing and applying radiferous luminescent products; researching or measuring radioactive substances or x-rays in laboratories; making radiotherapy or x-ray machines Jobs exposing workers to radiation in hospitals, sanatoriums, clinics, medical, dental, and radiological offices, oncological treatment centers and health institutions; Jobs in all industries or businesses using x-rays, radioactive substances, substances or devices that emit radiation as indicated above.

Tableau 6. Liste des maladies professionannelles indemnisables (in French)

Affections provoquées par les rayonnements ionisants

Date de création: 4 Janvier 1931 — Dernière mise à jour: 26 juin 1984 (décret du 22 juin 1984)

Désignation des maladies	Délai de prise en charge	Liste indicative des principaux travaux susceptibles de provoquer ces maladies
Anémie, leucopénie, thrombopénie ou syndome hémorragique consécutifs à une irradiation aiguë.	30 jours	Tous travaux exposant à l'action des rayons X ou des substances radioactives naturelles ou artificielles, ou à toute autre source d'émission corpusculaire,
Anémie, leucopénie, thrombopénie ou syndrome hémorragique consécutifs à une irradiation chronique.	1 an	notamment: Extraction et traitement des minerais radioactifs; Préparation des substances radioactives;
Blépharite ou conjonctivite.	7 jours	Préparation de produits chimiques et pharmaceutiques radioactifs;
Kéeratite.	1 an	Préparation et application de produits luminescents radifères;
Cataracte.	10 ans	Recherches ou mesures sur les substances radioactives et les
Radiodermites aiguë.	60 jours	rayons X dans les laboratoires; Fabrication d'appareils pour
Radiodermites chroniques.	10 ans	radiothérapie et d'appareils à rayons X;
Radio-épithélite aiguë des muqueuses.	60 jours	Travaux exposant les travailleurs au rayonnement dans les hôpitaux, les sanatoriums, les cliniques, les
Radiolésions chroniques des muqueuses.	5 ans	dispensaires, les cabinets médicaux, les cabinets dentaires et radiologiques, dans les maisons de
Radionécrose osseuse.	30 ans	santé et les centres anticancéreux; Travaux dans toutes les industries
Leucémies.	30 ans	ou commerces utilisant les rayons X, les substances radioactives, les
Cancer bronch-pulmonaire primitif par inhalation.	30 ans	substances ou dispositifs émettant les rayonnements indiqués ci-dessus.
Sarcome osseux.	50 ans	

A former maintenance worker promoted to a managerial position died of lung cancer in 1994 at the age of 52. He had worked exposed to radiation from 1965 to 1994 in a mechanics company operating at nuclear power plants. The lung cancer was discovered in October 1992. He resumed work from March to July 1993, continuing to move around among various plants. The company doctor filed a declaration of occupational disease in 1993. Recognition was granted after his death.

Research has already been done into the non-recognition of cancer as an occupational disease (Brugère, Thébaud-Mony, Pezerat, & Cassou, 1994; Réseau SCOP93, 2005; Thébaud-Mony, 1991, 1994). The French compensation system's restrictive approach to what diseases qualify for inclusion in Table 6 put an end to any hope Mme S. might have had of obtaining compensation. The doctor who wrote the original medical certificate underlined the diversity of radiation-induced cancers and stressed Monsieur S.'s exposure to ionizing radiation in the course of his occupational activity. But the compensation board kept to the letter of the Table 6 of the list of occupational diseases. The red tape and hardships of her life discouraged the young widow from pursuing her case. We must be sensitive to this other dimension of damage incurred: the fact that people find themselves up against a system that refuses to recognize the reality of an occupational illness and the extreme suffering it represents for the family of the deceased. Nothing can "repair" the loss of a companion and father, and posthumous compensation is experienced as an unbearable minimization of that suffering, bringing it down to a stingily calculated pension granted as a privilege instead of a right.

In the second case, of bronchial cancer, recognition of occupational illness came only after death. At the hour of his death the employee in question was deprived of what such recognition symbolizes, for himself and his dignity as a worker, and for his loved ones as the relief brought by posthumous financial support. Moreover, his cancer will not figure in CNAMTS statistics as due to energy production but to metallurgy, since this is the official activity of the company he worked for.

The particularities of DATR workers' occupational history mean there is little chance that their diseases will be recognized as occupational. Aside from difficulties in obtaining proof of exposure, their employment and tasks subject them to such a high degree of mobility that exposure sources are necessarily multiplied. For some workers, maintenance or industrial cleaning involves a variety of tasks that expose them to different sorts of carcinogenic elements (ionizing radiation, asbestos, soldering dust and smoke, benzene, other cancerous chemicals, etc.). The synergetic effects among these different products have not been documented in the medical or epidemiological literature and are thus not taken into consideration in the system of compensation for occupational disease.

Subcontracting thus allows for masking the ties that exist between illnesses due to radiation or maintenance work (even when the system recognizes these illnesses as legitimate motives for compensation) and the owner-operator EDF, which presides over the organization of work and exposure to the relevant risks. It therefore seems next to impossible to get closer to the realities of victims of the reduced life expectancy associated with the collective dose of ionizing radiation received in French nuclear power plants—other than by directly interviewing workers or their families.

According to the mortality statistics in Y. Tanaka's 1997 study of Japan, there were 6 times more deaths from cancer among "outside" nuclear industry workers than in the Japanese population at large.

Men and Women Made to Work Inappropriately or Excessively

This misuse of workers is generally related to work hours and schedules. Nearly half of employees participating in the STED survey have abnormal schedules (beyond legal work-week hours, and as high as 60 hours a week; some work 6 or even 7 days a week) and/or atypical hours (having to work non-standard hours, night or weekend shifts). Over 20% state they "occasionally or more often" have daily schedules exceeding 16 hours. One-third says they have irregular eating and sleeping schedules. Three out of four employees in the survey mentioned work pace constraints (urgent or intensive labor, frequent interruptions, periods of waiting around that seem to contradict the urgency of the work to be done).

There is a statistically significant correlation between these different types of time constraints and work intensification, affirmations of disturbed sleep, anxiety attacks and fatigue, as well as the perception of being insufficiently available for one's family. Moreover, for a segment of DATR workers, other anxiety-inducing factors, namely forms of financial and occupational insecurity, must be added to the difficult working conditions and constraints. It may reasonably be concluded that the combination of these circumstances favors road accidents.

Two employees in our study died in road accidents, in circumstances that could not be clarified:

- A 29-year-old inspector died in an accident in 1991. No information was forthcoming as to the circumstances.
- A nuclear servitude worker aged 30 died in an accident in 1992. The report communicated by the company doctor reads as follows:

> This employee was the victim of a fatal road accident in November 1989. The accident took place around 5pm as he was returning from work. There was no investigation by the CHS-CT. No connection would appear to exist between his working conditions and the cause of death. The patient had no particular medical history to speak of.

The French compensation board for occupational injuries recognizes such road accidents as having an occupational origin, even though for one of them the company doctor declared there was no formal connection between the accident and working conditions. The doctor did not indicate what facts this declaration was based on. What did he know of the working conditions, temporal constraints and nervous tension experienced by that worker on the day of the accident and the days preceding it?

Subcontracting employers cited in Part II stated that EDF demands made it impossible for them to comply with customary work scheduling regulations (despite the fact that scheduling is becoming increasingly more flexible). Once again, in the state occupational injuries and disease compensation system's statistics, mortal road accidents appear in the subcontracting company's activity sector, not in energy production. They therefore are not taken into account in assessments of the effects of nuclear industry work on workers' health.

Damage to Dignity and Self-Esteem

The results of a statistical survey do not easily allow for distinguishing between what is a matter of a person's individual, family, or group and collective history and what is more properly understood as experience of self and surroundings, interpersonal and social relationships. However, two points that emerged in interviews with DATR workers in the STED survey clearly suggest disturbance and damage in this latter area: nearly half the sample said they received no recognition whatever for their work from their superiors (this echoes accounts in Part I), and 33%, especially temporary workers, stated their intention to look for a different job.[12] Reasons given related to the hardships of moving around and not being available for family, which was frustrating for family life, as well as to financial and career considerations. The meaning of these answers only becomes fully accessible if we take into account what these workers say about their experience of daily life. Accounts in Part I referred to lack of appreciation for their work, the fact that family relations are negatively affected and, in some cases, the suffering induced by solitude.

A significant statistical correlation appeared in the STED survey between lack of appreciation for work performed or lack of time for family, on the one hand, sleep difficulties, fatigue and anxiety on the other.

Davezies analyzes this as an attack on personal dignity associated with work, leading to psychological suffering and especially "violence against oneself: exasperation is turned against oneself, and manifests itself in the form of pathogenic behavior and ultimately physical and mental illness. The most severe form that such self-directed violence can take is suicide." This goes some way to explaining the suicide of a 32-year-old temporary worker. He turned against himself the violence inherent in a socio-economic situation of financial

[12] According to the CRG Ecole Polytechnique survey, 82% of respondents wished to find another job (Dégot et al., 1999).

and job insecurity of the sort generated by the way the nuclear industry uses temporary workers.

The valve worker performing tasks that exposed him to radiation committed suicide in 1994 at the age of 32. When the company doctor who had seen him for the regular medical visit shortly before his death was questioned as part of the investigation, he gave the following answer:

> I asked X . . . (a temporary work agency), about the circumstances of the death of Monsieur P., 32 years old. I saw that employee in April 1994 on the occasion of a medical visit for DATR employment. I noticed nothing in particular. He had been followed before in the same service from 1986 to 1988 (6 visits) : there were no known pathological signs or treatment in his case.
>
> On the telephone, my correspondent (at Company X.) seemed to me extremely "relieved" to have the opportunity to speak of the circumstances of the death of Monsieur P., who had already worked for the company and whose work was appreciated there. Before this last job, Monsieur P. had been working for another company in the same activity sector. Company X had hired him on a limited-time contract. When these jobs were over, Monsieur P. found himself unemployed once again. He had confided to the temp agency director the financial difficulty he was in due to a mortgage. He had asked for a cash advance in order to have his automobile repaired.
>
> Concerning the circumstances of his suicide, this is what I was told : he went home and used a pellet gun to shoot himself in front of his wife and 15-day-old baby. The person who reported this to me is convinced that the immediate reason for his suicide was unemployment and its financial consequences.
>
> That is all the information I can provide about the circumstances of Monsieur P.'s death.

For Thomas (1993), suicide "constitutes a language that has to be deciphered." Nevertheless, the account given by the employer as reported by the company doctor quite directly identifies the cause as the employment and remuneration conditions the young father found himself caught in, precisely the conditions that ensure EDF the legitimacy it seeks for operating nuclear power plants and the formal criteria of radiation protection. That legitimacy is obtained at the cost of growing economic insecurity for "outside" workers, while EDF manages to lower average yearly individual dose.

The solitude and individualization of the situations experienced by temporary workers make the economic insecurity generated by such intermittent employment even harder to bear, keeping in mind that in accordance with the state unemployment insurance system's digressive unemployment compensation arrangements, payments diminish as the worker remains idle, causing a rapid and drastic loss of income. We are very far indeed from correct application the Code du Travail, which stipulates that no preventive measures for protecting worker's health must lead to any financial disadvantage for workers to whom they are applied (Code du Travail, art. L231-11).

In Great Britain, a widow managed to have her husband's suicide recognized as linked to his occupation, since it took place in a context of anxiety provoked by the real threat of being laid off.[13] In France, the first case of work-related suicide was recognized in 2005: the suicide was that of an EDF manager and psychological pressure was identified as a cause of death (Desbordes, 2006). Beyond the matter of compensating the person or family, the question arises as to possible strategies for preventing suicides in connection with occupational circumstances. In 1995, the Coordination of CGT Labor Unions at the Chinon nuclear plant learned of seven other suicides by "outside" DATR workers.

France has one of the highest levels of death by suicide in the world, and the rate is particularly high among low-skilled male workers aged 30 to 45. Surault underlines considerable social differentiation with regard to suicidal behavior, noting that "its mechanisms depend on the very way our society functions, the way it generates and reproduces economic and social inequality," and adding: "The specific policies aimed at preventing suicide can only be marginally effective given the inherent mechanisms of our social and economic organization" (Surault, 1997).

As in the previous cases, the death of Monsieur P. will never be counted as collective damage related to the organization of labor and working conditions that characterize maintenance operations in the nuclear industry. Victims of the risky, insecure conditions engendered by the forms of labor organization used in the nuclear industry thus disappear from appraisals of the "human cost" of nuclear energy production.

The Japanese study on mortality rates among nuclear industry "outside" employees also cites a high number of suicides.

Mental Suffering

Certain STED survey results gave rise to a theoretical construction in the field of labor psycho-dynamics (Molinier, Doniol-Shaw, Dessors, & Huez, 1996) that was then applied by Christophe Dejours in *Souffrance en France* (1998) as a system for interpreting the increasing instability and social precariousness of labor.

The results in question concern workers' interest in their jobs and their sense of responsibility and autonomy. Most DATR employees interviewed in the survey answered those questions positively. Molinier et al. posit that such answers are paradoxical and question their validity. This paradox is not put forward in the terms of a psychodynamics-of-labor approach with the same workers, an

[13] WHIN-Internet, March 6, 1998.

approach that aims to interpret the meaning of the STED survey's coded responses in terms of the subjective, qualitative expression of respondents giving those responses. Instead, the authors base their interpretation on results of other surveys, surveys where the respondents were not outside workers but EDF permanent employees, called *agents*, surveys bearing on EDF agents' opinion of subcontracted companies and the quality of their work at nuclear power plants (Molinier et al., 1996).

Those judgments cast doubt on the skills and know-how of subcontracted workers; also on the work relations that the latter claim to develop with each other and with EDF agents. And they posit uninteresting work content. Referring to these judgments, the authors of the article conclude that the work done by subcontracted workers "is not interesting." This leads them to evaluate outside DATR employees' responses to questions about responsibility, autonomy and interest in the job as merely a hollow "apology" with regard to the qualities of nuclear power plant work, an apology to be understood as due to the "prestige" of the nuclear industry; this in turn is understood to explain why those workers "overrate" the qualities they attribute to their occupational activity. The defensive ideology thus demonstrated is understood to open onto the many forms of suffering-related violence, and to deviant practices (drugs, gambling, brothels) and dissolution of family life among these workers.

Dejours returns to this theoretical construction in *Souffrance en France,* generalizing and systematizing it still further to include all subcontracting arrangements and all forms of unstable or precarious employment (Dejours, 1998). He draws a grim picture of sick workers tramping along European high roads, divorced, living in "trailers," and engaging in "marginal forms of sociability" that are looked upon disapprovingly by permanent EDF agents:

> This way of life, similar to that of nineteenth-century workers, inevitably leads to extra-normal social behavior use of alcohol and above all drugs, which temporarily calm workers' misery and despair. Prostitution is the ineluctable companion of disorganized morals and lifestyles. AIDS develops here more than elsewhere, and AIDS is frightening. It creates ghettoes at the very gate of the company.
>
> When the company's permanent staff, those in charge of supervising and overseeing these workers, actually come in contact with them, the workers provoke their suspicion, disgust, and even moral sanction. Indeed, it often happens that, because of their state, their work is plagued with mistakes, but also and above all, with fraudulent affirmations, due both to incompetence and lack of training, which they would like to hide, and to pressures and abuse on the part of the heads and directors of the subcontracting firms—pressure and abuse that cannot be controlled or made public. Thus these workers may cause production defects and have a detrimental effect on safety and security in spite of themselves, with dire consequences for permanent owner-operator employees. (Dejours, 1998, pp. 114-115)

This text calls into serious question the quality of subcontracted employees' work: because of their mistakes and concealments, they are understood to be responsible for incidents or accidents. To my knowledge there is no study attesting to frequent use of drugs or alcohol or a high incidence of AIDS among these workers. These ideas seem much more likely to be the projection of fears and perhaps fantasies haunting EDF agents with regard to these "outside" employees, keeping in mind that EDF agents experience the subcontracting of maintenance as dispossessing *them* of their own know-how and skills in nuclear installation maintenance.[14] These declarations contradict our results. Applying a psychodynamics-of-labor approach with these workers would of course give us different knowledge from what sociological or epidemiological approaches provide, and also allow for interdisciplinary exchange. But shouldn't such an approach be used within the framework of a direct survey of workers themselves about the "dark side" of nuclear energy production?

How Will They Age?

The various ways that health is negatively affected by work organization and working conditions in nuclear power plants depend, in some cases, on knowledge acquired over decades. But the imprecise nature of these processes prevents them from becoming explicit issues in either compensation for occupational disease or disease prevention. This is especially the case for relations between health and work hours and schedules. Citing the contradiction between the prescribed temporal framework and "worker's own time," Laville brings to light "the serious consequences" that this contradiction causes for "workers' physical, mental, psychological, and social health" (Laville, 1984). STED survey data concerning the work hours and schedules of outside DATR employees performing maintenance tasks in the nuclear industry bear out the hypothesis of damage to health, damage that Laville defines as "traces that gradually restrict personal, professional and social life and activity." The average age of the first survey population in 1988 was 40. What will become of aging workers who can no longer bear the working conditions observed? Recent research on the age of working populations shows how, in stable working groups, labor is "redivided" and tasks "redistributed" in relation to age and experience (Laville & Volkoff, 1993; Marquié, Volkoff, & Paumes, 1996). The type of division of labor introduced by

[14] In a 1996 letter to the authors of the article and the journal that published it, F. Guillon, M. Lallier, and I initiated a discussion of this interpretation, which is stigmatizing for "outside" workers and based on weak results, taken from a survey that can hardly be said to have used a psychological or psychodynamics-of-labor approach (see Guillon, Lallier, & Thébaud-Mony, "Précarisation du travail, sous-traitance et santé: quelles recherches pour quelles transformations? Discussion-commentaire à propos de l'article 'Sous-traitance et exclusion sociale' de Molinier et al." Letter to the authors, November 12, 1996).

subcontracting has done away with the necessary leeway for such reorganization, reorganization that would reduce exclusion of aging workers. Continuous turnover among DATR workers, who are constantly being replaced by younger ones free of a "radioactive past," spares the nuclear industry the costly restructuring of the "labor force" needed to ensure maintenance of nuclear installations.

CONCLUSION

The fact that there are no contract ties between EDF and "outside" workers conceals EDF's responsibility for cases of accident, sickness and death connected to work done while exposed to radiation in nuclear power plants, as well as its responsibility for exclusion of workers. The social and sexual division of labor in French society—"men without families, women without work" (see *Le sexe du travail*, 1984)—confers social legitimacy on this type of labor organization. And this form of labor organization in turn confirms the invisibility of families who are suffering from the death of husband, father, provider, and of the labor exclusion of men who, for the owner-operator EDF, amount to no more than labor power bought from other employers. Calculating kilowatt-hour cost is a way of ignoring the victims and the harm done to those they leave behind. Yet workers' deaths and the occupational factors related to them show that there are and will be more victims than suggested by the ICRP's specific-risk assessment.

Given that we are prevented from imputing risks and their consequences to those who are responsible for their reproduction—i.e., nuclear industry owner-operators—are we going to be faced with a situation similar to the case of asbestos? Are we going to have to see other victims, victims considered in social terms to be more worthy of attention than subcontracted workers—i.e., men and women exposed daily to the effects of "natural" radioactive contamination, contamination which is increasing due to the diverse and largely unsuspected ways that risk is spreading—or even undergo a more serious nuclear "incident" than the others, an accident leading to radioactive contamination of great magnitude reaching beyond the limits of the nuclear power plant itself, before effective prevention measures are taken?

CHAPTER 7

Safety of Nuclear Facilities or Obliteration of "Traces"?

> EDF: nous vous devons plus que la lumière!
> [We owe you more than the light!][1]

General safety is the keystone of French social and political consensus about the use of nuclear power. According to the directors of France's nuclear installations, general safety is the central principle for determining how nuclear energy production and plant maintenance are organized and managed. However, the safety authorities (DSIN) and the parliamentary committee (Office Parlementaire d'Evaluation des Choix Scientifiques et Technologiques), that both surveys safety and radiation protection in France, have been expressing growing concern about the frequency of "incidents" in basic nuclear facilities and increasing instances of malfunction in radiation protection (see the committee's annual reports for the 1990s). The fact is that safety and radiation protection are directly influenced by how maintenance work is organized.

This chapter works to demonstrate the hypothesis that the strategies used to circumvent the obstacles created by the two major contradictions facing nuclear industry managers imperil the safety not only of EDF and subcontracted workers but also of the civilian population beyond power plant gates. The absolute refusal to recognize the activity of "outside" workers or give them any opportunity to speak out on their working conditions in nuclear plants works to obliterate the "evidence" or "traces" of how safety is being threatened.

Starting with the observation of reduced safety margins in nuclear installations, an observation that applies to several western countries, this chapter looks at some of the incidents that have occurred in French plants during maintenance operations, thereby shedding light on, and validating, the above hypothesis. Lastly, I consider the consequences of "disorganization of work," in Dwyer's sense of the term (1991), in light of the frequency and nature of these incidents.

[1] EDF advertising slogan disseminated in the media.

REDUCED SAFETY MARGINS

Serious nuclear accidents have drawn analysts' attention to the conditions that imperil nuclear safety. Some of the points identified by these observers correspond to what worries safety authorities in various countries.

Industrial disasters such as those that occurred at Three Mile Island, Chernobyl, Bhopal, and Seveso, in industries priding themselves on high reliability, have in common a chain of problems that were not understood as problems and were thus left untreated and unsolved. At Three Mile Island and Chernobyl in particular, warning signals given by operators before the accidents occurred went unheeded. It was as if the confidence that high-level technicians and risk management specialists put in theoretical models and prescribed procedures had distracted them from reality. If reality doesn't correspond to the model, is it 'the real" that is mistaken, or should the model be rehauled in accordance with what the real reveals? The Russian physicist Legassov, a designer of the Soviet nuclear program, commented in his suicide letter of April 26, 1988 (the second anniversary of the Chernobyl accident):

> The information I had led me to think that all was not well in the development of nuclear energy. . . . Routine was gradually gaining the upper hand, routine where work was concerned but also in problem-solving. I realized all this, but it was hard for me to get involved; my general statements on the subject were not at all appreciated because any attempt by a non-specialist to get a hearing for his ideas on the work of scientific organizations was considered unacceptable. . . . This explains how a generation of engineers grew up who, though skilled of course, lacked a critical attitude in matters of equipment and systems for guaranteeing equipment and facility safety. (quoted in Belbeoch & Belbeoch, 1993)

Legassov thus points up certain elements of the social construction of nuclear accidents, and the weight in that construction of the absolute power of the "generation of uncritical engineers," inattentive to warning signs that would have allowed for anticipating the accident and perhaps modifying the course of the conditions that made it possible.

In several countries, inspections or audits have brought to light organizational malfunctioning in the nuclear industry linked to the practice of subcontracting. In Canada and Great Britain, audits of nuclear facility safety brought to the fore the danger to safety caused by subcontracting and the use of temporary workers. In Canada, the report commissioned by the executive managers of the Canadian company Ontario Hydronuclear and prepared by the IIPA/SSFI group (Independent Integrated Performance Assessment/Safety System Functioning Inspection), which developed out of the Nuclear Performance Advisory Group, was particularly critical of how production and maintenance on Ontario nuclear sites were managed. The expert evaluators saw organizational choices themselves and their effects on operators' work activity as the very causes of "unacceptable erosion of safety margins" in the Ontario nuclear industry (IIPA/SSFI, 1997) Among other problems, the report mentions productivity pressure from and on managers; an exceedingly high proportion of temporary contracts; insufficient preventive maintenance for ensuring long-term facility reliability; the refusal to allow operators a forum for speaking about their work and excessively slow modification of work procedures that needed changing; extensive radioactive contamination of certain sites; absence of plans for handling emergencies. The recommendations were aimed not so much at changing operators' behavior as bringing about necessary changes in organization and management.

In Great Britain, a 1999 report by the Health and Safety Executive criticized the way work was organized at British Nuclear Fuels and the company strategy of cutting worker numbers, which had led to the development of practices that ran counter to safety. By reasoning exclusively with the aim of lowering costs, affirms the report, the firm engaged so many subcontracting companies that it lost control of day-to-day plant operation. Changes were imperative, and failure to make them would incur sanctions against the company (HSE, 1999).

The extremely serious accident that occurred in October 1999 in the Japan Nuclear Fuels Concession Company, responsible for manufacturing nuclear fuel, likewise involved threatening safety by outsourcing labor, all in the aim of reducing costs. Instead of following a mechanical procedure for controlling the masses of uranium in nitric acid solution, the company had its personnel work with their bare hands and metal buckets to transfer liquified uranium to the decantation tank. There was no particular protection or containment, nor any automatic procedure for halting the process of filling the decantation tank. Sixteen kg of uranium were transferred into the tank instead of 2.4 kg, thus causing a nuclear reaction—what is known as a criticity accident,[2] the most dreaded of those likely to occur given the uncertainties in the nuclear industry. A high level of gamma and neutron rays was emitted, and radioactive fission products were released into the air of this highly populated region 100 km from Tokyo. Three workers received an extremely high, potentially fatal dose, and two died. The study brought to light that these were temporary workers who had not received adequate training. It was also discovered that the company had two sets of operation manuals: the first was submitted for inspections and was in keeping with official public norms; the second contained the instructions that had really been given to employees. Clearly the primary concern was profitability, to the detriment of safety. An entire region was irradiated (Citizens' Nuclear Information Center, 1999).

In France, the DSIN safety authority of the Ministry of Industry assesses "incidents" that have occurred in nuclear power plants—several hundred a

[2]Criticity refers to the moment that critical mass is reached and certain radioactive elements set off a nuclear reaction.

year—in its annual report. For 1998 the number was 376, more than one a day. A. C. Lacoste's diagnosis of the situation at EDF in 1998 was particularly critical (DSIN, 1998):

> A great deal of news in 1998—some of it unsettling. Though the safety of the population was never endangered and the number of incidents actually decreased from 1997 to 1998, numerous events received a great deal of media and/or technical attention, which helped bring to light the following causes of malfunction:
>
> - negligence clearly caused the affair of surface contamination of train cars transporting spent nuclear fuel . . . ;
> - problems encountered on the Belleville plant enclosure were due in part to workers' falling asleep . . . ;
> - a lax attitude was one cause of failure in a shut-down cooling circuit in Civaux I on March 12. . . .;
> - at least twice in the course of the November 27, 1998, Golfech incident that contaminated approximately ten service company employees (fortunately not very seriously), very simple things had been forgotten. . . .
>
> Alerted by a mobile alarm beacon that had just discerned an abnormal level of radioactivity, the plant's radiation protection service decided to check the information given by the beacon before evacuating the thirty subcontracted company employees working at that level of the building. This slowed evacuation of those workers by more than an hour. Clearly, things should have been done in the opposite order.
>
> Contamination on one of the work sites was due to use of a filtering fan equipped with a defective filter. The investigation showed that this filtering system had not been inspected since it was put in place. . . . . (Excerpts from introduction to report summary, Lacoste, January 28, 1999)

It should be noted that this warning from the safety authority to the directors of France's nuclear installations does not go beyond hypotheses about behavior: negligence, lack of maintenance, lax attitude, forgetting simple things. It does not develop a critique of work organization itself as related to safety issues.

THE SOCIAL DIVISION OF LABOR AND RADIATION DOSE IMPERILS SAFETY

The fact is that the "disorganization" of work by subcontracting and the social division of labor and radiation dose imperil safety by entailing a loss of practical knowledge of real work.

"Outside" workers have no forum or space in which to give direct expression to their experience. Their knowledge of nuclear facilities and the mastery this knowledge gives them of what occurs in nuclear plants does not endow them with any legitimacy to issue alerts, estimate a risk, warn about the fragility of a given structure or the inappropriateness of a given procedure or use of a particular instrument. The extreme dispersion of operators brings about an equivalent dispersion of this precious capital of knowledge and experience of the condition of equipment and facilities on the various nuclear sites. After a season of unit shutdowns during which operators have moved from site to site from week to week, they no longer remember the operations performed on this or that valve, on this or that reactor in a given plant. The written traceability required by EDF cannot replace living memory of facilities when it comes to ensuring safety. Numerous studies in sociology, ergonomics, and the psychology of work show that what workers have to say about the reality of work situations opens up understanding of what is functioning properly and what is not in production processes and the work procedures imposed on operators.

Moreover, the current rigid, formal type of operation and organization tends to make workers less likely to use the maneuvering room they have or to anticipate problems, whereas those two practices would help to ensure that operations are performed in better conditions (they would allow workers to make better use of their know-how and the practices associated with a particular job specialization). The groups that develop in the course of regular work, and that ensure the quality and dependability of operations *by means of informal collective regulations,* are constantly being dissolved by the uncertainties and chance involved in subcontracting.

The fact that an individual is sanctioned for getting irradiated or contaminated (wage reduction, job loss, warning) makes concealment of real work particularly likely because it gives "outside" workers an incentive to camouflage instances of radiation exposure and to abandon the dosimetric measuring and recording instrument. To this must be added the absence of any dosimetry procedure focused on individual operations. It becomes impossible to identify work posts where it is difficult if not impossible to diminish radiation emissions or set up a technical individual system of protection. This means that there is no way of making visible the places, conditions, and time-frames where contradictions between safety and radiation protection for workers as it pertains to maintenance cannot be resolved.

It thus becomes impossible to assess in exact terms how extensive low-level radioactive contamination is inside or outside nuclear plants and how much intermittent irradiation a worker population operating in a great number of sectors and dispersed throughout the national territory is receiving.[3] The type of work organization chosen by nuclear industry principals engenders dissemination of radioactivity risk in a way similar to the problem for asbestos. But there is a major difference between the two. Whereas asbestos is a fiber that can be located materially, nuclear risk is impalpable. The absence of reference to any means of measuring this risk other than individually received radiation dose means that

[3] It should be recalled that "outside" workers receive 80% of the collective radiation dose.

debates on the effects of this danger ignore the reality of actual exposure. Internal contamination or high-level exposure that occur in an extremely brief lapse of time are not declared as work accidents,[4] and in the medium or long term, the division of labor through subcontracting is an obstacle to coherent medical surveillance for this population, surveillance of the sort that would allow for identifying those who are going to develop radiation-caused cancer and for declaring and getting this cancer recognized as an occupational disease. In all these cases, the actor that produces the danger—the owner-operator EDF—has figured out how to not be accountable for the consequences; i.e., how not to compensate outside workers for work accidents or occupational disease. This it has accomplished through the form of work organization it uses.

Directors of France's nuclear installations can of course take pride in having lowered the average annual individual dose received, but only at the cost of permanently obliterating the traces that would be brought to light by analysis of work situations and listening to what outside workers have to say about both their activity and possible signs of a decline in their health.[5] To this must be added the effects of still less visible phenomena; i.e., dispersion of radioactive contamination outside the nuclear plant, for which the nuclear industry is also responsible, as is shown in analyses of "incidents" that have occurred in France in recent years.

INCIDENTS REVEAL CONTRADICTIONS

It is important to distinguish between two types of "incidents": those leading to the dissemination of radioactivity at and outside nuclear plant sites, and those that could lead to a full-blown nuclear accident.

The first category includes some of the incidents that have irradiated or contaminated "outside" employees and—much less often—EDF agents operating on installations during unit shutdowns, as well as incidents leading to radioactive pollution outside the plants. It can happen that workers leave the site without being totally decontaminated. When they are called back for inspection the next day, they must leave all personal effects and belongings that have been contaminated by radioactive particles (including sheets); these are then disposed

[4]Chemical poisoning, HIV contamination, or hepatitis C are declared as work accidents, thus giving the victims the right to recognition and compensation should deferred effects appear. Given that irradiation and contamination (which security definitions qualify as incidents) are particular events occurring at a given moment in work activity, they are very similar to poisoning. In this sense they are indeed work accidents and not occupational diseases, which refer instead to the notion of "usual exposure to risk."

[5]Doctor-work inspectors seeking to improve medical surveillance of outside DATR workers through a survey of industrial medical service personnel and the setting up of systematic recording of incidents in a work accidents log in nuclear power plants ran up against opposition from INB company doctors, who accused them of intensifying "the present confusion in debates about the effects of low-dose radiation."

of as radioactive refuse. Environmental pollution occurs when nuclear waste or spent fuel is transported (as in the incident of the contaminated Valogne train cars)[6] or when radioactive steam or water leaks outside plant enclosures. These incidents constitute so many safety breaches because they permanently disseminate radioactivity in the public space, either by irradiation and contamination of workers themselves or by low-level pollution, which has been spreading in recent years.[7]

The second type of incident involves instances of unanticipated technical or organizational malfunction the scope of which is itself unpredictable. The human error argument generally allows the principal EDF to conceal any relation between the incident and work organization, specifically the way maintenance operations are organized. As Dwyer (1991) has shown, "incidents" can be considered the "product" of social relations in the workplace as those relations are organized by subcontracting. Here we look at incidents that occurred between 1995 and 1999 on different nuclear sites in the framework of maintenance operations during unit shutdowns. In some cases it was possible to interview "outside" workers who were present when the incident occurred. For obvious confidentiality reasons and because these workers could be fired if the incidents in question were identified, I shall not describe them individually.[8]

[6]"Following a surveillance visit conducted jointly by the Direction de la Sûreté des Installations Nucléaires and the OPRI on April 28, 1997, at the Valogne railway terminus (this station is the property of COGEMA La Hague nuclear reprocessing plant), it was observed that 35% of the train cars used by EDF to transport spent nuclear fuel to La Hague showed surface contamination at at least one spot that was above the regulation limit of 4 becquerels per cm^2. Of the 44 EDF train cars that had been used in 1997, 10 showed external contamination. At this stage, however, there is no reason to think that there has been any health risk for the population at large. Investigations are under way. The DSIN has called on EDF to explain the origins of this contamination and the corrective measures that have been taken (press communiqué, OPRI and DSIN, April 30, 1998).

[7]In 1997, when the contamination of nuclear waste transport train cars leaving nuclear power plants and heading for La Hague was discovered, the EDF and COGEMA owner-operators had known for ten years that the cars used to transport radioactive waste were already contaminated when they left the plants.

[8]When I requested assistance from OPRI services and company doctors in interviewing "outside" workers involved in a contamination incident affecting employees of several service companies (I had received official approval for the research project), OPRI answered that I had to go through the doctors working for the service companies, and that I should ask the EDF company doctor at the relevant INB to put me in contact with those doctors. The company doctor refused for the following reasons: he had "not participated in drawing up the research project" I showed him, and he had "no desire to draw media attention to the incident, which would revive worries whose medical-psychological impact is difficult to control, not so much because of the incident itself or even its technical or health consequences but with reference to the sensational effect the incident had outside the site."

Level of Compensation

This refers to the issues of remuneration for and recognition of work.

The preceding chapters showed that the wages of outside company employees are low compared to those of civil-servant status permanent EDF agents. The main issue, however, is uncertainty about maintaining income at a given level, not only because of the threat of total or partial unemployment but also the danger of radiation exposure. Pay level is not maintained if worker reaches the maximum dose (though it should be according to labor regulations); it is instead reduced as soon as the employee reaches the "peg," regardless of his contract (see chapter 3). This means that labor relations around the matter of paying employees in case of major exposure to radiation turn compensation into a kind of punishment or sanction—as if the worker were himself responsible for (and guilty of) getting the dose he has received. In an incident, this process is intensified in three directions.

- At the level of employment, warnings and even threats of being fired followed upon contamination incidents, and the concerned workers were not even given a chance to explain how the incident occurred, either to EDF officials or their superiors in their own companies.
- At the symbolic level, there is loss of personal effects or belongings—a watch or chain, for example, gift from or reminder of a loved one—but above all there is the sanction of being immediately demoted and disqualified (a team head or project manager being demoted to mere underling, for example), and there is an absolute refusal to recognize the strategies used by these workers to limit the scope of the incident and even to prevent the most serious of its consequences.
- At the level of legal action, "outside" workers have been found guilty in some incidents. In the case of a radiography operation where solder seams had not been checked as they should have been, a criminal proceeding led to finding the team of operators guilty without taking into account the entire set of facts, which showed that the way the operation had been planned made it absolutely impossible to perform the set of verifications called for. Unexpected cracking had forced the operators to do the necessary repair work, and therefore to spend most of the time allotted for the whole operation handling this chance occurrence. "Manage!" was the only instruction that the team of operators received. Subjected to two imperatives—one from their own company, which did not want to be penalized for not meeting EDF's deadline, and one from EDF, which refused to envisage any kind of change in the work schedule—they had no choice but to submit and carry out repairs on selected solder seams that they had been able to check before the deadline. This strategy, which was reported to EDF officials as fraudulent behavior on the part of the outside company by another employee who had been laid off was judged in a criminal court, and the outside workers were found guilty.

The "Command" Level

This concerns the level and forms of power exercised by management over workers in organizing and running work, and it play a major role in accidents. It is at this level that individual and collective worker maneuvering room and resistance are created or destroyed. It is also at this level that workers can either have or be deprived of the capacity of showing and making heard their own knowledge of work situations—knowledge constructed through real activity, involving real work—and the possibility of getting that knowledge taken into account in matters of work organization and accident prevention measures. Dwyer (1991) has shown that the more tyrannical the type of authority and power, the more frequently work accidents are "produced." Forms of "voluntary servitude" are said to come into play when constraint is imposed as an obligation to produce results (without any negotiation of the means for doing so) or the threat of being laid off or fired. The consensus between employers and employees about non-compliance with safety arrangements and rules is particularly relevant to this form of command in the company, especially in subcontracting. Still, as Dwyer points out, only workers are actually subjected in their very bodies to the effects of a lack of safety arrangements.

At this level, the social relation that develops is one of absolute power exercised by nuclear plant managers over "outside" workers with regard to prescriptions for work, deadlines and procedures, "return on experience" modes, interpretation of facts related to incidents, and information about incidents (as much information given to the media as to the public authorities).

Incidents often reveal this absolute power. Several of the incidents that have been analyzed have been shown to be linked to errors in prescription (mistaken valve number noted in the operation file; indication of one circuit in place of another; valve left open that should have been closed; iodine filter used instead of a total filter) or an error in reading or interpreting a prescription (confusing two valves with similar numbers; operating on a functioning circuit). In both of these cases, the "human error" was said to have been committed by "outside" workers operating at the time of the incident, and those workers were given no chance to defend themselves against arbitrarily meted out sanctions—unjust in the case of a prescription error, and always unwarranted when considered in relation to what should have been done to prevent this type of incident. Indeed, these "errors" are most often indicative of malfunction at the level of organization: a confusing way of numbering valves; too many worksites to be surveyed at one time by an overworked EDF agent; poor knowledge of the site among operators with temporary contracts; excessive time pressure. Taking such malfunction into account implies criticizing the type of organization that caused it.

In other cases, the work schedule was maintained without any possible objections, even when chance occurrences (the discovery of cracking in a given length of piping, worn-out seals that had to be changed though they were scheduled to hold for a longer time) made it impossible to accomplish the tasks in the fixed amount of time. In some situations, the fact that it was impossible for "outside" employees to negotiate other operation strategies led to irradiation incidents (for example, during decontamination of a tank: using a robot beforehand would have spared workers from receiving high levels of radiation).

The owner-operator EDF, citing its formal procedures of "written traceability," has the power to exclude any consideration of real work from the process of assessing work situations in which incidents have occurred. Dosimeters and x-ray films that measure radioactivity "disappear," and it becomes impossible for the concerned worker to gain acknowledgment of the dose he says he got against claims from radiation protection services indicating a different number. The task of a cleaner is changed into verification that there are no leaks on a given circuit, thus making him responsible for a delay in evacuating the contaminated worksite. According to the worker's own account, the leak, which he had signaled earlier to the relevant EDF departments, had not elicited any reaction. Any unplanned event occurring in nuclear installations during unit shutdowns tends to be considered by EDF agents as the "fault" of outside workers.

In several of the incidents studied, warnings given by subcontracted workers first led to EDF's checking that the warning was justified. This caused a delay of several hours before the order to evacuate workers in a contaminated atmosphere was given. But the press communiqué issued by plant managers blamed the "service company" workers for this delay, claiming they had not complied with the safety instructions calling for immediate evacuation of a contaminated worksite. In one situation studied, there were cases of internal contamination; outside workers had ingested or inhaled radioactive particles. These instances of contamination did not lead to any declaration of a work accident.[9] The special CHS-CT meetings called on union demand after incidents do not include representatives of the implicated personnel or service companies. In most cases, CHS-CT delegates do not receive any name-specific information on the involved "outside" workers. They therefore have no access to the workers' version of the facts. Without the testimony of the actors present in the installations at the moment of the incident, the work as it had been prescribed takes the upper hand over real work in reconstructing and interpreting the facts.

The immediate dispersion of outside workers present during incidents seems a non-written rule imposed by EDF on service companies, a rule applied

[9]During a CHS-CT special meeting called after an incident in which several outside workers had undergone internal contamination, a company doctor declared that "getting internally contaminated is not an accident to be handled by the legislators as long as the degree of contamination remains within the authorized limits—the same criteria as for external irradiation." This definition of the work accident represents an extremely loose interpretation of the Code du Travail, which defines work accidents as "any accident occurring on work premises or during work" without placing any limit on exposure.

immediately, i.e., the very day after the incident. Not only are operators not asked to explain what happened, but also their bosses are warned or even called in, and the workers themselves are sometimes transferred immediately to another plant, far from the one in which the incident occurred. Officials in charge of the site decree the official interpretation; all employees (including EDF agents) are forbidden to speak off-site of what, in their opinion, occurred. The official interpretation carries three messages, aimed at safety and radiation protection authorities and the public at large (through the media). First message: the incident was caused by "human error." "An untimely act," "unawareness or inattentiveness," "non-compliance with safety rules" are some of the standard expressions in communiqués on the circumstances of the incident. The second message is that no nuclear pollutants were released into the environment or if some were, the amount was below the authorized level. The third message is that radiation or contamination levels for individual workers remain below official quarterly or annual limits. Obviously it is not specified that the dose was received very fast. If the radiation dose actually exceeded the limits, the message changes, and dose received is said to be below the level at which health is observed to be adversely affected. This refers to direct effects observable a few days or months after the incident, since, as explained in chapter 6, regulation limits are used to determine a "socially acceptable" increase in cancer among exposed populations but not to determine an effects threshold. These messages conceal or make it difficult to ascertain the chance occurrences or problems that actually led to the incident. And they gloss over any link between the incident and possible health consequences for the persons who were irradiated or contaminated.[10]

The Organizational Level

This refers to the types of division of labor and distribution of tasks imposed by management. It is at this level that the constantly maintained split between "noble" work—design, oversight—and disparaged work characterized by routine—execution—comes into effect. It is also at this level that the various types of "disorganization" of work which workers must adapt to in their activity—i.e., changes in production, such as the adoption of just-in-time methods or different types of flexibility (of employment, work hours, work itself)—occur. Citing his own empirical studies of situations in France and New Zealand as well as books

[10] In the weeks and months following a contamination "incident" on the site of an owner-operator other than EDF, a temporary worker presented a number of symptoms directly listed in the literature (extreme fatigue, hair loss, affected lymph nodes, saliva loss, intestinal and esophagus problems). Despite two operations on the lymph nodes, he was unable to get the "incident" recognized as a work accident, an accident whose consequences should have led to significant compensation (100% coverage of treatment costs, increased daily benefits, recognition of permanent partial incapacity).

and publications on trends in national work accident statistics, Dwyer (1991) showed that most accidents in what are called advanced industrial societies are associated with this organizational level. Organization is imposed on workers in the form of structural constraints that dispossess them of control over the sources of danger in the tasks they are called upon to perform. Dwyer hypothesizes that the heavier the weight of this organizational level as imposed on workers—in relation to the dangers in their tasks—the greater the risk of accident.

In the nuclear industry, it is at this level that the risks are highest, because organizing maintenance by means of subcontracting and unstable employment entirely transfers the matter of managing the very real contradictions involved in maintenance from the directors of France's nuclear installations at large and site directors to "outside" workers in the framework of their operations during unit shutdowns. Analysis of incidents brings to light the difficulties encountered by workers in performing their tasks—despite all—while limiting as much as possible the number and scope of incidents linked to "disorganization" of work engendered by flexibility and the division of labor and radiation dose.

On the basis of the cases studied, we can hypothesize that if no major accident has occurred up until now in the area of maintenance, this is not only because safety defense procedures have been in the main complied with, but above all because applying these procedures—together with occasionally transgressing them—is coherent with the intelligence, know-how, and sense of responsibility characterizing operators in the field, whether these be outside workers or EDF agents. Without minimizing the importance of the latter, it is important to underline that the former are constantly demonstrating their sense of responsibility, by accepting—in the name of safety but also the urgent economic imperative that nuclear plants be able to function at top capacity as much of the time as possible—temporal and material working conditions that imperil their own health. This is the "voluntary servitude" Dwyer spoke of, but it includes a strong reference to public service as represented by the kind of energy production their activity contributes to. This may be illustrated by the following facts.

At the level of worksites, contradictions may be observed between the teams of the various companies with regard to task that each is to perform and its needs. Several workers cited problems as basic as not enough electric plugs for the needs of the various teams. In one case, some operators on a given site were in *Mururoa* suits while others had much more rudimentary protection; the latter got contaminated. The tasks in question should have been done successively rather than simultaneously, but because of the work schedule, the teams operated at the same time.

Fragmenting and spreading the dose among tasks requiring a high level of exposure prevents "outside" workers from performing an operation to the end and "taking their time." Having to follow each other in time segments of a few minutes while the dosimeter is busy measuring degree of radiation that one is being subjected to does not help individual workers remain in control of what they are doing. This contradiction between safety (quality of the operation) and radiation protection is often pointed up in outside workers' accounts of radiation and contamination incidents. But it never figures in the official reports on these incidents.

The business agreement between the principal and the subcontracted companies means that operations have to produce results whatever the chance occurrences. This gives rise to time constraints that become in themselves a danger to safety: working seven days a week; staying in the reactor building more than 20 hours running to finish an urgent job; soldering or machining components of structures that "spit" because decontamination was cut short in order to make the spot available for the next set of operators. Incident frequency is consistent with the organizational risk-taking represented by just-in-time methods. The pressure to produce results means that outside workers become responsible for chance or unplanned events, for alarms that go off (or fail to). A nuclear servitude company team leader working at night, doing the final inspection of a site before turning it over to the next team, discovers major leakage of radioactive water. No alarm goes off and he is the only one who is aware of the problem. What should he do? The rule is categorical: he is to alert the relevant EDF department. He does so, but nothing happens. Should he wait? Should he begin an emergency decontamination process? Or immediately evacuate his co-workers? How is he to measure the danger? What will the consequences be to himself if he interrupts work? Will he receive a warning? Will he be fired for gross misconduct? For deserting the site? What will the consequences be for his company? Lateness penalties? Loss of future jobs? Faced with this situation, this team leader made choices that incurred sanctions against him.

One last extremely important dimension of the organizational level pertains to types of union representation. As mentioned above, "outside" workers are not represented on CHS-CTs at nuclear power plants. In case of incident, they have no representative to defend their word and the understanding of the facts for which they are being faulted. Understanding the relations of domination that exist between EDF agents and outside workers, and how readily EDF agents become suspicious about the work of outside employees, the logic of actions performed by those workers often escapes EDF agents and is not mentioned in their testimony. This in turn means that union representatives are themselves party to the division of labor created by subcontracting; they too end up supporting the technicist and technocratic safety ideology according to which "We're protected by the experts." On the other hand, outside workers, who know how incidents happen and to what degree their work is implicated in safety, also know that the "disorganization" of work and working conditions brought about by subcontracting has profound negative effects on their individual and collective maneuvering room in work activity. Repression of attempts to create union branches in nuclear service companies deprives outside workers of any space of collective expression in which the reality of conditions that make incidents and accidents possible—a reality closely linked to subcontracting—could offer resistance to the "human error" ideology. Is it going to take a major serious accident to stop this headlong flight from reality?

HEADLONG FLIGHT: HOW FAR AND WITH WHAT CONSEQUENCES?

We may illustrate the contradictions that agents must accept in their work activity by a serious radiation accident and the reactions to it—at the time it was considered merely an "incident." On March 13, 1999, a brief article in *Le Monde* indicated that an EDF technician had been irradiated. According to the EDF communiqué quoted in the newspaper, the victim was in charge of "checking, without having to enter [into the red zone], that a set of spotlights that had been temporarily installed in a zone adjacent to the reactor tank had been taken out again. When he saw that they were still there, he took the initiative of removing them." The first communiqué indicated that the agent "had received a dose of approximately 80 mSv—over the regulation limit of 50 mSv per year." The newspaper *Libération* of the same date explained that the dose was not 80 but 340 mSv, and that the DSIN had decided to submit the case to the courts for investigation. Then the weekly magazine *L'Evénement du jeudi* dated March 25-31, 1999, published an investigation by the journalist Desbordes entitled "Tricastin: did the radiation victim prevent a disaster?" In an interview, the irradiated agent is quoted as saying, "I have orders, sir. I can't talk." Questioning the various actors involved and reading EDF's official communiqués led the journalist to wonder: "The communiqués say nothing about the motives that led the employee to intervene as he did. What was the nature of the risk? Did the technician's intervention prevent a much more serious accident? Did he choose to 'sacrifice himself' when he went in to remove the two spotlights? The fact is that he acted for a reason. An internal dispatch of the *Agence d'Information du Tricastin* sheds some light: 'For the technician, the two spotlights might have exploded or imploded when the area was put under pressure. What would have happened in that case? When he became aware of an imminent accident, the technician chose to intervene.'"[11]

But nothing is said about what would make it possible to better understand the sequence of events. Nor is anything said of what occurred before the incident—the operations performed during the unit shutdown, namely the "disorganization" of work—that would allow for understanding how and why the spotlights necessary for those maintenance operations came to be "forgotten" there. The contradictions remain hidden from prying eyes. . . .

[11] *L'Evénement du Jeudi,* March 25-31, 1999.

An IPSN report underlines the importance of knowing what went on *before* incidents with regard to safety. Referring to the results of a 1992-93 ergonomic study by the IPSN and the ergonomics department of the *Conservatoire des Arts et Métiers* as well as on her own experience of the work activity of service company employees, acquired through safety studies done by IPSN in EDF power plants, Baumont analyzed the relations between "human factors, the way unit shutdowns were organized, and safety issues." The study is aimed at clarifying "the compromises actors make to take into account the entire set of determinants in an shutdown (deadlines, workers safety, operation conditions) and in particular those affecting safety." Analyzing the regulation process that ensures that safety issues will emerge and be handled, Baumont situates the issue of transmitting information at the very heart of this process (Baumont, 1999):

> Real safety assessments are constructed on the basis of information from references substantiated by information on the real state of the facility. The construction of this body of information is *a job in itself* [author's italics] involving the different actors who were operating during the shutdown. This work presupposes seeking information, constructing a representation, and treating the safety issues that arise.

She stresses that

> contrary to information from theorical references, the information that emerges in the process of the work only exists to the degree that it is the result of construction in live time, either because it was officially sought so that a trace of operations could be left, or because someone takes the initiative to circulate it and thereby to submit it to the judgment of other persons.

She shows that the narrow margins of the regulation process result primarily from "limited circulation of information about the real state of things" (see box p. 226), and concludes:

> To get beyond the various limitations, it is important to create an organizational context that will facilitate communication between service operators and EDF agents. The state of the facility after the maintenance period is in part due to information, information transmitted by a population external to the facility and therefore absent during the period it is in use. This information is relayed by a small number of persons having a direct view onto the material (inspectors and project heads ["chargés d'affaire"]; there is no trace of it in the documents if it is not reported there.
>
> In conclusion, the time, the means and the skills necessary for informal regulation are still insufficiently organized, and there is still no real reflection on the tools and methods that might increase regulation process efficiency. (Baumont, 1999, pp. 63-64; author's italics)

Baumont's study brings out a number of structural limitations in regulation and in the handling of safety questions, limitations that arise from these choices. However, the ergonomic approach does not go so far as to examine issues of power and social relations, which clearly have a major impact in the situation related in the preceding paragraphs.

Limited Availability of Information

The information that can be collected on the site of the operation is often not well described with office words; the result is that information is lost between the work site and the report on the work that is made in the offices.

Information transmission time can enter into competition with finishing the work day (for service company employees) or the work week (for project heads) and with the many other problems that they have to handle simultaneously, namely writing up official documents.

The tools (computer applications, plans, and other documents) that are made available do not always allow for transferring and circulating the information sought by different actors; the search for information is not facilitated by the computer applications made available for this purpose.

At every level, each person makes a hypothesis as to what is relevant and should be transmitted. For unexpected information with safety implications to reach persons in charge of making safety-related decisions presupposes that all the links in the chain have perceived that the information is important to safety.

Actors prefer to keep the information they have about their own activity to themselves in order to maintain a certain power or mastery over their work and to avoid possible sanctions against themselves or their company should they be suspected of being responsible for events.

Lastly, actors may be induced to stop transmitting certain information if they feel that the information they transmitted the last time was not taken into account. The quality of information transmission depends on the effectiveness of the "return on experience" in the way information is organized.

The last, most complex limitation is as follows: it can never be taken for granted that the question that would enable people to realize that an action during an operation is likely to lead to a safety problem will actually be raised by one of the partners involved in the operation" (Baumont, 1999, pp. 62-63)

The way maintenance work is organized and the social relations underlying this form of organization favor an increase in the risk not only of incidents but also of accidents. But this type of organization also makes it possible to obliterate all signs or traces of the major contradictions that the nuclear industry is confronted with. Industrial history and certain of the disasters that have occurred in the course of it are there to remind us that the contradictions always end up exploding into the light of day. They force societies to confront them and overcome them. Do we have to have a nuclear accident in France in order for the nuclear power lobby, the state, and the society at large to take on these contradictions? The fact is that even for such a scenario, the nuclear industry has everything planned; i.e., it has already gotten special exemption from public accountability laws: in October 1968 a law was passed limiting the financial accountability of the nuclear owner-operator company to 50 million francs [less than € 8 million]. Above and beyond that sum, victims are to be compensated by the state.

CONCLUSION

The very substance of this book shows that the "traces" of work activity can never be entirely obliterated. The reality of the contradictions in the nuclear industry becomes manifest within its own confines through the frequency of incidents—incidents are the visible tip of the iceberg. The invisible part is what only "outside" workers experience, workers on the side of real work, the work of doing maintenance in nuclear facilities. Only if these workers are given the individual and collective right to speak out about what they experience in their work, only if decision-makers listen attentively to them, can we attain knowledge of the risks perpetuated by these contradictions and hope to find effective means of controlling them.

Conclusion

Subcontracting maintenance work is the means that the nuclear industry has chosen to make it seem that it is possible to reconcile two contradictory aims. Those contradictory aims are ensuring installation safety and the safety of workers responsible for the human labor that is indispensable to running those installations; and keeping nuclear energy competitive by making maximum use of nuclear facilities while ensuring that the maintenance tasks required to produce nuclear energy are performed during minimum-length unit shutdown periods. Through a longitudinal survey in the form of interviews with "outside" workers in the French nuclear industry and a study of the social organization of labor and radiation protection, this book has sought to present the social conditions in which this work is done and their impact on the health of these workers and nuclear safety at large. The major obstacle to our research was the social invisibility of this composite "underside" of nuclear energy production. The study nonetheless allows for identifying what is at issue for public health, and indicating future research avenues that should be given priority.

WORK IS AT THE CORE OF NUCLEAR ENERGY PRODUCTION

Far from having "disappeared" as the dominant discourse today would have us believe, work—in the sense of physical, psychological and manual labor—is at the very center of the issues raised by nuclear energy production and nuclear safety. The activity of nuclear industry workers, both EDF agents and "outside" company employees, is indispensable to the functioning of nuclear installations and "possesses a specificity comparable to none" (Jobert, 1998). In "Contribution à unc anthropologie de la reconnaissance au travail," Jobert brings to light the lack of work recognition for EDF reactor operators, non-recognition profoundly characteristic of labor in the nuclear industry. The specificity of that labor is its deep implication in the most fundamental of human individual and collective concerns: life and death. For Jobert, "the contribution [made by reactor operators], a contribution which cannot be recognized through the logic of reciprocity, is to take on and perform for society at large an activity that carries in it both life and death" (Jobert, 1998, p. 219). Nuclear energy brings life in the form of electricity, which has become indispensable to life today and to the renewal of life; it brings death through the danger that cannot be dissociated from it, and the terrifying use of it represented by the bombs dropped on Hiroshima and Nagasaki. Jobert underlines the historical and symbolic importance of this characteristic in the constructed representations of power and might that we have attached to the nuclear industry: "Like the public at large, the generation of researchers, engineers, and owner-operators working in both the military and civilian nuclear sectors all belong to the Hiroshima generation. This means that from a very early age and well before they knew what occupation they would choose, they have all been acculturated to a Manichean vision of nuclear energy" (Jobert, p. 231). In Jobert's anthropological analysis, the relation between reactor operators and the EDF company that employs them is understood to involve an exchange of numerous "gifts" and "counter-gifts" that go far beyond mere wage-earning and paying arrangements and that weave strong ties between the two categories. But this reciprocity dynamic does not answer the need for recognition experienced by reactor operators.

The maintenance function is of course not the same as the reactor operation function; maintenance is rather the indissociable underside of running reactors. Reactor operators cannot practice their skill unless that skill is "secured" (as we speak of being "secured" in mountain-climbing) by the skill of the "outside" workers who maintain nuclear installations, who repair them when necessary and who labor to prevent breakdowns. Maintenance work too draws its meaning from the contribution made by workers "external" to this "life and death-bearing" activity, for it is an activity that all nuclear industry workers assume responsibility for. These workers commit their physical person. They take on the unavoidable danger of their activity—the danger of losing health and life—so that others—*all* others: EDF agents and society at large—are protected from deadly exposure to ionizing radiation. But this investment, this commitment receives no recognition. In contrast to the situation of EDF reactor operators, maintenance workers operate in conditions that break radically with the reciprocity dynamic. Not only are wages low for most of them given the skill they invest in the work, but also wage levels are constantly being pulled lower. The economic security of "outside" workers is constantly being threatened; it is permanently "precarious" in the literal sense of that word. Moreover, obliteration of the traces of these workers' real work as brought to light by this book also constantly causes their skill, and the responsibility they assume as a public service they have been entrusted to perform, to disappear from memory of nuclear energy production work. These workers know that they practice an activity essential to electricity production. But the absence of any formalized relation of reciprocity between them and the EDF Company—namely the fact that there is no work contract—puts them in a social position of subordination and condemns their work, and the risks they are subjected to, to invisibility. It is in this way that the illusion of no-risk nuclear energy is sustained for French society as a whole.

PUBLIC HEALTH ISSUES

Two major public health issues are at the core of the social organization of maintenance work in the nuclear industry. First, the threat to implementation of effective occupational health policy that is represented by the increasing instability and precariousness of employment and social protection; second, the matter of how to handle major technological risk.

The Increasing Instability of Work as Related to the Health of "Outside" Workers in the Nuclear Industry

The employment stability that EDF agents enjoy and their low level of exposure to ionizing radiation are made possible by a strategy of transferring highly exposed maintenance work to outside workers. The use of subcontracting, sub-subcontracting, and temp work in connection with the nature of tasks to be performed enables nuclear energy manufacturers to divide up radiation dose among outside workers and thus ensure compliance with dosimetric norms. Subcontracting creates a particular relation of domination between the principal and employees who are not its own, a relation in which the work contract no longer plays its role of "cornerstone of the juridical definition of the work relation" (Supiot, 1994). This means that the employee no longer has the maneuvering room that would allow him to withdraw from dangerous situations. Norms concerning only one of the risks incurred are substituted for the general risk prevention principles that constitute the foundation of business owners' obligation to ensure safety. The extreme formalism of compliance with these norms conceals the deleterious effects on the health, employment, work and living conditions of these outside workers. It also conceals violation of a number of clauses in the Code du Travail concerning the use of temporary workers, work scheduling, the obligation to maintain wages even when safety measures are adopted, and the obligation not to discriminate against workers in terms of health. To this must be added the invisibility of work-related health damage and of principal's civil and penal responsibility for it. Furthermore, compliance with dosimetric norms does not exhaust the obligations incumbent upon nuclear energy manufacturers in the matter of health in the workplace. Those obligations pertain to respect for the bodily integrity of "outside" workers operating on their installations.[1]

The results of this study bring to light the substantial impact of risk of exposure to ionizing radiation on the individual and collective organization of work in nuclear power plants, and the impact of the other work modes and constraints discussed on individual and family health. The three-pronged flexibility—flexibility of jobs, work hours, and family life—has serious and sometimes tragic effects on the health and equilibrium of workers themselves and their families. The death of five members of the surveyed population is the cruelest manifestation of this.

In our report to the Ministry of Labor (Thébaud-Mony et al., 1992), when the effects of subcontracting on work relations and the threat to the health of outside workers was becoming clear, recommendations were made by the working group coordinating the preliminary stages of the research. In particular, it appeared absolutely necessary to prohibit sub-subcontracting in the interests of public health. Moreover, the numerous instances of regulation-bending suggest an urgent need to develop more rigorous arrangements for verifying that occupational health legislation is being applied and to make principals, i.e., nuclear plant constructors and owner- operators, more accountable for work-related health damage.

The results of the present study are consistent with the generalized research observation that working conditions throughout the European Union have worsened and that with current intensification and flexibility of labor, work organization constraints have greatly increased. The changes under way in "the world of work" (title of Kergoat, Boutet, Jacot, & Linhart, 1998) amount to a regression of labor regulation and a serious threat to the health of workers in France and the European Union. Clearly more attention needs to be paid to the conditions of health in the workplace in determining employment and work-schedule policies.

[1]Article L225-14 of the French penal code should be recalled here: "Taking advantage of an individual's vulnerable or dependent position in order to subject him or her to working or living conditions that are incompatible with human dignity is punishable by 2 years of imprisonment and a 5000-franc fine." Two other clauses in the penal code directly apply to health in the workplace. One pertains to involuntarily endangering a person's life (Art. 222-19) and involuntarily causing physical damage to a person (Art. 221-6). In the latter case, "deliberately failing to respect a safety obligation imposed by law or regulation" constitutes an aggravating circumstance. The other pertains to causing risk to others: "directly exposing others to an immediate risk of death, mutilating injury, or permanent infirmity through manifestly deliberate violation of a particular safety or precaution obligation imposed by law or regulation" (Art. 223-1) is punishable by imprisonment.

Handling a Major Risk

The forms taken by individualized management of exposure to radiation make the nature and specificity of these cases of exposure and their immediate or delayed effects socially invisible. This way of managing risk is part of a general development in occupational health research and policy-making. In a study of scientific and occupational practices related to the design and use of genetic testing for selecting workers, Draper has critically analyzed the scientific paradigms underlying these practices (Draper, 1991), showing that between 1970 and 1990 a shift occurred from an approach in which occupational risks and dangers were studied *together* with working conditions and dangers to one in which the focus of observation has shrunk to workers themselves as "risky" individuals.

In the nuclear industry, the development of radiosensitivity research has shifted scientists' attention from collective conditions of exposure to individuals. Priority goes to molecular biology and genetic epidemiology over any other scientific approach. There are no toxicology studies or clinical research on the immediate or delayed effects of ionizing radiation as it occurs in the specific conditions that "outside" employees work in. This observation may be extended to epidemiology: the electronuclear program has been in existence for more than forty years in France and epidemiological surveillance of the most exposed occupational population—outside workers—has not yet been developed. To the present day there is no program in place for recording and studying the symptoms and pathologies affecting this population, which as we know gets exposed to far more radiation in French nuclear installations than any other group.

The radiation effects reference model still being used concerns a radically different event from the occupational exposure analyzed here. That model was developed from the results of epidemiological observation of Japanese populations exposed to radiation from atomic bombs during World War II. For constructing knowledge and developing understanding of the pathogenic effects of occupational exposure to radiation, this constitutes a real blind spot, one that will be difficult to remedy given the extraordinary dispersion of exposed persons, not only in France but in all countries that have adopted the same model for organizing maintenance work in nuclear power plants. In the struggle against diseases associated with ionizing radiation, this is a wrong perpetrated against both present and future generations.

With regard to risk prevention, throughout this study I have identified numerous contradictions between the training and oversight strategies developed by nuclear power manufacturers and the increasing instability and precariousness of employment and health for employees of outside companies. A highly specialized prevention policy aimed at formal compliance with individual exposure limits is counter-productive for people's health if it is not embedded in an overarching effort to improve work conditions. In his report on how the European Union framing directive of 1989 on organization of risk prevention in the workplace was being applied, Vogel (1994) makes the following points:

> While certification guaranteeing correct compliance with procedures instituted to ensure health and safety can produce useful information, it will never allow us to take into account the complexity of real work and the many, diverse individual and collective defense strategies. Integrating issues of health and well-being at work into quality control procedures would therefore make sense as a complementary policy instrument. But it would be illusory to make this the main thrust of prevention policy, and if it led to systematic claims of "human error," it could have dangerous consequences for workers if the work did not correspond entirely to prescriptions for it.

The set of training and licensing procedures set up by nuclear energy manufacturers partakes of this narrow conception of prevention and of business self-regulation in the crucial matters of general and worker safety. There is no concern for the development of individual and collective knowledge, nor is there any use of workers' own work- and health-related experience. Clearly, the disappearance of any negotiated form of wage regulation between work prescriber and those who perform the work also profoundly undermines the form and content of risk prevention.

The Undermining of Nuclear Safety?

How far can the rationalization logic of cutting costs go without imperiling nuclear safety? Subcontracting of maintenance work leads to the dispersion of practical experience for which no "written traceability" can substitute, and in so doing deprives industrial societies of knowledge of the work that is absolutely necessary to nuclear energy production. By bringing to light the practical experience that outside workers possess, the survey presented here shows the knowledge we are beginning to lose and the processes of socially constructing incidents and accidents to be associated with factors of "work disorganization" and the social relations that cause it. Nuclear installation safety is threatened by these social processes in two ways: the danger of radioactivity being diffused into the world situated beyond leak-proof reactor circuits and plant enclosures; and the danger so clearly attested to by nuclear incidents that one such incident, more serious than the preceding ones, will become a major accident.

With reference to the public health issues of this situation, can we speak of "residual risk" becoming socially acceptable through a cost-benefit calculation? With the increasing development of dismantling, transportation, storage and reprocessing operations, tons and tons of radioactive waste are going to be handled. This issue goes beyond French borders. What will the effects be on "client" countries of exporting nuclear power plants, sold "key in hand," particularly on countries where workers have infinitely fewer labor rights than in France? The contradictions inherent in nuclear energy production are the same everywhere. The conditions in which upkeep and maintenance are performed are thus likely to bring about the socially invisible irradiation and contamination of thousands of workers in these countries, as well as chronic or accidental radioactive pollution of the environment.

RECOMMENDATIONS FOR NEW RESEARCH

Several types of research work are necessary—in the social sciences, occupational health, and the life sciences—to make up for the knowledge deficit brought to light by the results of this study.

There has been little social science research on subcontracted work, and what there is often done not on the basis of direct surveys of subcontracted workers but rather with permanent employees and employers in the companies who use these workers. To gain systematized knowledge of how maintenance work is organized in the nuclear industry and the impact of such organization on working conditions and health, an approach that combines economics, sociology, legal studies, and ergonomics is essential. This research should consist in a prospective study of a unit shutdown that would cover all stages, from how principal determines which companies to subcontract to completion of the various jobs, and include a precise listing of workers operating on the site while recording their work itinerary during the shutdown period and before and after it. Clearly, the researchers would need to be present on the worksites. The research would also entail detailed identification and analysis of all incidents occurring during the shutdown, to be done on the basis of accounts from the various actors involved in those incidents, beginning with the workers who experienced them directly.

Occupational health research should be integrated into the framework of organized, permanent surveillance of workers exposed to ionizing radiation. The regulation framework needed for such a study is provided by the Code du Travail, which requires medical surveillance for persons exposed to cancer-causing substances. But given the current conditions of employment instability, the work of organizing that surveillance has yet to be done. It requires taking into account the serious malfunctioning observed in the preliminary stages of this research, in order to radically transform the way occupational health services are organized. The first concern is ability to ensure identification and medical surveillance of all workers (including self-employed ones). EDF should pay for this surveillance, which should guarantee workers a right to consistent, coherent medical follow-up and, if necessary, recognition of occupational disease. The second concern is to develop hypotheses about health damage among this population and to validate them in the framework of epidemiological protocols that take into account the precise conditions in which workers are exposed to risks—that of ionizing radiation but also other types of risks and possible synergies between them—by work post, task performed, type of exposure, type of radiation or other pollutant, and so forth. This approach would allow for critical discussion of current dominant models in the debate on effects of low doses, discussion based on detailed research data on occupational exposure.

In cancer epidemiology, latency periods—but also the intense instability of employment and the invisibility of health damage that only manifests itself years and even decades later—seriously undermine the validity of all current epidemiological approaches based on surveys of stable, weakly irradiated populations and using average radiation exposure values. Just as in occupational epidemiology we cannot extrapolate risks from Hiroshima and Nagasaki results alone, so the work and exposure conditions of permanent EDF agents and subcontracted employees differ too greatly from those of outside workers to allow for extrapolating risks for the latter from results of research conducted with the former. New research must be developed that will allow for studying the risks that "outside" workers are subjected to in terms of the specific characteristics of their work and employment conditions.

Lastly, in life sciences, the specificity of these working conditions could generate hypotheses for new toxicology and biology studies. For low radiation dose in particular, experimental research should allow for comparing effects of the various types of exposure that occur in maintenance activity: low-level continuous exposure, intermittent but intense exposure, different types of radiation (alpha, beta, gamma, neutron), and radiation in synergy with other toxic materials (asbestos, chemical substances). This would be a way of testing health effects of the various low radiation dose models. It would allow for looking critically at the validity of those models, not only theoretically but also in reference to the reality of observed instances of exposure. Hasn't the use of average values been a means of obscuring the need to develop new knowledge necessary for protecting and preserving public health?

The nuclear industry has been a field for trying out new forms of work organization and risk management. New research instruments on work and health are necessary to rigorously assess the health effects of these new forms of organization and management.

The proposed research projects are particularly necessary given the fact that the nuclear industry is moving into the phase of dismantling installations. From an economic perspective, this is a situation similar to the one for asbestos. After a long period in which asbestos was at the center of a lucrative productive process, it has now become a costly waste material that cannot be destroyed and must therefore be managed, removed, stored. The same is true for the spent fuel and contaminated structures of nuclear power plants. It is going to take centuries to deal with these materials. What kind of public health policy, what kind of institutional vigilance, will allow us to effectively protect the health of the "temporary and subcontracted" operators who will be performing—for French society as a whole—the tasks required for managing this waste?

Bibliography

Académie de médecine, Académie des sciences. (2005). *La relation dose-effet et l'estimation des effets cancérogènes des faibles doses de rayonnements ionisants.*, Rapport d'un groupe de travail mixte, Paris.

Académie des sciences. (1995). *Problèmes liés aux effets des faibles doses des radiations ionisantes.* Rapport de l'Académie des Sciences No. 34, octobre.

AEN. (1995). *Tchernobyl. Dix ans déjà. Impact radiologique et sanitaire.* Paris: OCDE.

Alexievitch, S. (1998). *Voices from Chernobyl: The oral history of a nuclear disaster.* (Keith Gessen, trans.). Dalkey Archive Press 2005.

Altershon, C. (1992). *De la sous-traitance au partenariat industriel.* Paris, L'Harmattan.

Appay, B. (1993). Individu et collectif: Questions à la sociologie du travail et des professions. L'autonomie contrôlée. *Cahiers du GEDISST*, No. 6.

Appay, B., & Thébaud-Mony, A. (dir, 1997). *Précarisation sociale, travail et santé.* Paris: IRESCO.

Ashmore, J. P., Krewski, D., Zielinski, J. M., Jiang, H., Semenciw, R., & Band, P. R. (1998). First analysis of mortality and occupational radiation exposure based on the national dose registry of Canada. *American Journal of Epidemiology, 148*(6), 1998.

Autorité de Sûreté Nucléaire. (1995). *Le contrôle de la sûreté nucléaire en France.* Paris: DSIN.

Autorité de Sûreté Nucléaire. (1999). *Rapport d'activité 1998.* Paris, DSIN

Aven. (2006). *Enquête auprès des Vétérans : conséquences sur la santé des essais nucléaires français. Résultats au 12/01/2005* (Website : http://www.aven.org)

Aven. (2006). *Enquête auprès des Vétérans. Conséquences sur la santé des essais nucléaires français. Résultats sur 1500 questionnaires.* Lyon, 2006

Barillot, B. (2003). *Les irradiés de la république. Les victimes des essais nucléaires français prennent la parole.* Coédition GRIP–Editions Complexe, Bruxelles–Lyon.

Baumont, G. (1999). *Facteurs humains et organisation des arrêts de tranche: Enjeux pour la sûreté.* IPSN, Rapport DES 363

Beaud, S., & Pialoux, M. (1999). *Retour sur la condition ouvriére.* Fayard, Paris.

Becker, K. (1998, February). Low-dose cost/benefit assessment—A view from Europe. *Health Physics, 74*(2).

Belbeoch, B., & Belbeoch, R. (1993). *Tchernobyl, une catastrophe.* Quelques éléments pour un bilan Editions Allia, Paris.

Belbeoch, R. (1988). Effets cancérigènes des faibles doses de rayonnements. In: GSIEN/CRIIRAD, Santé et rayonnement, *la Gazette nucléaire*, No. 56/57.

Belbeoch, R. (1998). Comment sommes-nous "protégés" contre le rayonnement? Les normes internationales de radio protection. *Le rôle de la Commission internationale de Protection Radiologique.* Paris, document non publié.

Bendazhersky, Y. (2005). *La philosophie de ma vie. Journal de prison. Tchernobyl: 20 ans aprés*. Paris: J. C. Gawsewitch.

Berneau, T. (1997). *Etudier l'optimisation de la maintenance conditionnelle sur le CNPE Chinon*. EDF-Production-Transport.

Bertell, R. (1985). *No immediate danger: Prognosis for a radioactive earth.* Summertown, TN: Book Publishing Company.

Bilan de l'application du décret du 29 novembre 1977 fixant les prescriptions particulières d'hygiène et sécurité applicables aux travaux effectués dans un établissement par une entreprise extérieure. Ministère du Travail, de l'Emploi et de la Formation professionnelle, Paris, octobre 1986.

Birraux, C. (1992). Radioprotection et installations nucléaires: Réformer pour plus d'efficacité et de transparence. *Radioprotection, 27*(3).

Birraux, C. (1993/1996). Contrôle de la sûreté et de la sécurité des installations nucléaires. Office parlementaire d'évaluation des choix scientifiques et technologiques. Assemblée nationale rapport No. 1008, Sénat No. 280.

Birraux, C. (1994). *Rapport sur le controle de la sureté et de la scurité des installations nucléaires.* Office parlementaire d'évaluation des choix scientifiques et technologiques, Assemblée Nationale No 1008, 470p.

Birraux, C. (1997). Contrôle de la sûreté et de la sécurité des installations nucléaires. Office parlementaire d'évaluation des choix scientifiques et technologiques. Assemblée nationale rapport No. 3491, Sénat No. 300.

Blank, V. L. G., Andersson, R., Lindén, A., & Nilsson, B. C. (1995). Hidden accidents rates and patterns in the Swedish mining industry due to involvment of contractor workers. *Safety Science, 21,* 23-35.

Bourdieu, P. (1979). *La distinction. Critique sociale du jugement.* Paris, ed. de Minuit.

Bourdieu, P. (sous la direction de) (1993). *La misère du monde.* Paris: Seuil.

Bourrier, M. (1999). *Le nucléaire à l'épreuve de l'organisation.* Paris: PUF, le Travail Humain.

Boutet, J. (dir., 1995). *Paroles au travail.* Paris: l'Harmattan.

Boyer, R. (dir.) (1986). *La flexibilité du travail en Europe.* Paris: La Découverte.

Brenner, D. J. (2006). In: Controverse: Les faibles doses de radiations ionisantes sont-elles carcinogéniques? *Bulletin Epidémiologique Hebdomadaire,* no. 15-16/2006.

Bretin, H. (1994). *Santé des ouvriers du nettoyage á Montréal et á Paris. La face cachée du Travail dans la ville.* ISIS/INSERM; MIRE, PARIS.

Broda, J. (1976). Problèmatique de la sous-traitance et du travail temporaire, analyse d'un cas: La zone de Fos et le système Solmer. Thèse de 3e cycle. Université d'Aix en Provence.

Brugère, J., Thébaud-Mony, A., Pezerat, H., & Cassou, B. (1994). Cancers professionnels: Connaissance, réparation et prévention. *Bulletin Cancer, 81,* 14-21.

BTS/SALTSA. (2000). Le Travail sans limite? Conférence Européenne, Bruxelles, Septembre.

Busby, C. C., & Yablokov, A. V. (2006). Chernobyl: 20 years on. Health effects of the Chernobyl accident Documents of the ECRR 2006 No. 1.

Caire, G. (1990). Le travail atypique salarié en France. In: G. et J. Rodgers (dir.) Les emplois précaires dans la régulation du marché du travail. La croissance du travail atypique en Europe. Genève, IIES.

Callon, M., & Latour, B. (1991). *La science telle qu'elle se fait.* Paris: La Découverte.

Canguilhem, G. (1966). The normal and the pathological, translated by Carolyn R. Fawcett in collaboration with Robert S. Cohen; with an introduction by Michel Foucault. New York: Zone Books.

Cardis, E., & Esteve, J. (1990). Etude internationale concertée sur le risque de cancer chez les travailleurs de l'industrie nucléaire. Protocole de l'étude de faisabilité. Rapport interne du Centre international de Recherche sur le Cancer No. 90/001A.

Cardis, E., & Esteve, J. (1992). Etude internationale concertée sur le risque de cancer chez les travailleurs de l'industrie nucléaire. Protocole. Rapport interne du Centre International de Recherche sur le Cancer No. 92/001

Cardis, E., Howe, G., Ron, E., Bebeshko, V., Bogdanova, T., Bouville, A., et al., (2005, July 9). Risk of cancer after low doses of ionising radiation: Repospective cohort study in 15 countries. *British Medical Journal, 331*(7508), 77.

Cardis, E., Gilbert, E. S., Carpenter, L., Howe, G., Kato, I., Amstrong, B. K., et al. (1995). Effects of low doses and low dose rates of external ionizing radiation: Cancer mortality among nuclear industry workers in three countries. *Radiation Research, 142,* pp. 117-132.

Castel, R. (1995). *Les métamorphoses de la question sociale. Une chronique du salariat.* Paris, Fayard.

Chanteur, J. (1992). *Dispositions réglementaires nationales. In: Colloque sur la Surveillance de l'exposition individuelle,* Paris, 17/18 octobre 1980.

Charte de Progrès établie entre Electricité de France et les organisations professionnelles représentatives des entreprises prestataires de l'exploitation du parc nucléaire. *Compagnie Corporate,* janvier 1997.

Chateauraynaud, F., & Torny, D. (1999). *Les sombres précurseurs: une sociologie pragmatique de l'alerte et du risque.* Paris: Editions de l'EHESS.

Citizens' Nuclear Information Center. (1999). *Worst ever accident in japanese history was an accident waiting to happen*. Tokyo: Internet Report.

Clarke, P. (1996). La controverse du seuil. *Radioprotection, 31*(4).

Clot, Y. (1995). *Le travail sans l'homme? Pour une psychologie des milieux de travail et de vie.* Paris, La Découverte.

Clot, Y., Rochex, J. Y., & Schwartz, Y. (1990). *Les caprices du flux. Les mutations technologiques du point de vue de ceux qui les vivent.* Paris: Ed. Matrice Mire.

CNAMTS. (1987/1993). *Statistiques financières et technologiques des accidents du travail.* Paris.

Comité d'Information des Professions de Santé. (2005). Epidémiologie. Etude internationale par cancer des travailleurs de l'industrie nucléaire. *La Lettre Nucléaire et Santé Actualités* No. 51.

Comité de protection Radiologique et de santé publique. (1998). *Evolution de radiobiologie et de radiopathologie: Répercussions pour la radioprotection.* Paris: AEN-OCDE.

Comité de radioprotection, groupe de coordination en radioprotection EDF. (1993). *Livre blanc de la radioprotection. La radio protection à EDF: Orientations et objectifs.* (Imprimé par Jouve, Saint Denis, janvier 1994).

Comités mixtes à la production Conseil Supérieur consultatif EDF/GDF. (2002). *Le nucléaire et l'homme. Le rôle et la place du travail des et des hommes dans la qualité et la sûreté nucléaire. Comité mixte à la production. Actes du Colloque,* Paris.

Commission Déniel. (1997). *Rapport de la commission instituée par l'article 30 de la Loi No. 96-1160 du 27 décembre 1996 de financement de la sécurité sociale pour 1997.* Ministère de l'Emploi et de la Solidarité.

Cottereau, A. (1983). L'usure au travail : interrogations et refoulements. *Le Mouvement Social,* juillet-septembre, No. 124.

Cour des comptes. (2002). *La gestion du risque accident du travail et maladies professionnelles.* Rapport au Président de la République suivi des réponses des administrations et des organismes intéressés, février 2002.

Cristofari, M. F. (1997). Les accidents du travail, indicateurs de précarisation de la santé au travail. In: B. Appay & A. Thébaud-Mony (dir, 1997) *Précarisation sociale travail et santé.* Paris: IRESCO.

Cru, D. (1998). *Note de lexture sur "Souffrance en France, la banalisation de l'injustice sociale" de C. Dejours, Seuil, 1998; Chimères,* 3e trimestre 1998.

CSSIN. (1995). *Rapport du groupe de travail sur la radio protection des travailleurs des entreprises extérieures.* Paris.

DARES. (1993). Conditions, organisation du travail et nouvelles technologies en 1991. Dossiers Statistiques du travail et de l'emploi. N°90-91-92, juin; les horaires de travail en 1991. *Dossiers Statistiques du travail et de l'emploi.* No. 98-99, octobre.

Dassa, S., & Maillard, D. (1996). Exigences de qualité et nouvelles formes d'aliénation. *Actes de la Recherche en Sciences Sociales,* No. 115, Décembre.

Daubas-Letourneux, V., & Thébaud-Mony, A. (2002). *Work organization and health at work.* http://www.fr.eurofound.eu.int/publications/EF0206.htm

Davezies, P. (1995). Des inégalités. *Revue santé et Travail.* No. 12 (juillet 1995).

Davezies, P. (1997). Processus de précarisation, organisation du travail, santé mentale. In: B. Appay & A. Thébaud-Mony (Eds.), *Précarisation sociale, travail et santé.* Paris: IRESCO.

Davezies, P. (1998). L'organisation et les transformations du travail à la lumière des atteintes à la santé. In: *Les Assises pour l'Amélioration des Conditions de Travail.* Revue des comités d'Entreprise et Organismes Consultatifs CGT. No. 77, juin 1998.

Dégot, V., Globokar, T., & Girin, J. (1999). *Les intervenants de maintenance nucléaire: Compétences, conditions de vie et de travail, attachement au milieu.* Rapport d'Etude du CRG de l'Ecole Polytechnique de Paris.

Dejours, C. (1992). Pathologie de la communication, situation de travail et espace public. Le cas du nucléaire. *Raisons Pratiques, 3.*

Dejours, C. (1995). *Le facteur humain. Que sais-je?* Paris: PUF.

Dejours, C. (1998). *Souffrance en France. La banalisation de l'injustice sociale.* Paris: Seuil.

Desbordes, J. P. (2006). *Atomic parc. A la recherche des victimes du nucléaire.* Paris: Actes Sud.

Dessors, D., Schram, J., & Volkoff, S. (1991). Du "handicap de situation" à la sélection d'exclusion: une étude des conditions de travail antérieures aux licenciements économiques. *Travail et Emploi, 48,* 31-48.

Dessors, D., & Volkoff, S. (1992). Le travail comme modeleur: processus d'adaptation et mécanismes de sélection. *Future Antérieur,* 16.

De Terssac, G. (1992). *Autonomie dans le travail.* Paris: PUF.

Doll, R. (1997, December). Effects of small doses of ionizing radiation on human health. *Nuclear Energy, 36*(6).

Dodier, N. (1995). *Les hommes et les machines.* Paris: Métaillé.

Doniol-Shaw, G. (1993). Sous-traitance et maintenance dans les centrales nucléaires. *Travail,* No. 28.

Doniol-Shaw, G., Derriennick, F., & Poitiers, F. (1998). *La dosimétrie des salariés de la sous-traitance nucléaire. Exploitation des données de l'enquête STED de 1993.* Rapport de recherche RR 9804 GDS, 1e phase, LATTS-CNRS, Marne la Vallée.

Doniol-Shaw, G., Huez, D., & Sandret, N. (1995). *Les intermittents du nucléaire.* Octares ed., Toulouse.

Doniol-Shaw, G., Huez, D., & Sandret, N. (2001). *Les maux de la sous-traitance. Enquête STED 1993-1998. Suivi sur 5 ans des salariés de la sous-traitance nucléaire.* Octares Editions, Toulouse.

Dossier "Sous-traitance." In: *Revue Travail,* No. 28, Printemps 1993.

Draper, E. (1991). *Risky business. Genetic testing and exclusionary practices in the hazardous workplace.* Cambridge: Cambridge University Press.

Druck, M. G. *Terceirização: Desfordizando A Fábrica—Um estudo do Complexo Petro-químico.* 02. ed. Salvador/São Paulo: Edufba/Boitempo, 1 [a] ed. (1999), 1[a] reimpressão (2001), 1999. 280 p.

Duclos, L., & Mauchamp, N. (1994). *Bilan-perspectives des relations sociales et prodes-sionnelles à EDF-GDF.* Rapport de Recherche. GIP Mutations industrielles—LSCI.

Durkheim, E. (1930). *Le suicide.* Paris: PUF (10e tirage, 1986).

Dwyer, T. (1991). *Life and death at work. Industrial accidents as a case of socialy produced error.* New York and London: Plenum Press.

ECRR. (2003). *2003 recommandations on radiation risk.* Regulator's Edition, Brussels.

ECRR. (2006). *Chernobyl: 20 years on. Health effects of the Chernobyl accident.* C. C. Busy & A. V. Yablokov (Eds.). Wales: Cardiff.

EDF/EPN Département communication. (1996). *Argumentaire CIPR 6O. Note de synthèse,* Paris.

Exploitation du parc nucléaire. *Rapports d'activité, EDF,* Les Editions Stratégiques, Paris, 1996.

Exploitation du parc nucléaire. *Rapports d'activité, EDF,* Les Editions Stratégiques, Paris, 1997.

Fargeot, C. (1997). Interim, reconquête et grandes manoeuvres. *Liaisons Sociales-Le Mensuel,* Juin 1997.

Forman, D., Cook-Mozaffari, P., Darby, S., Davey, G., Stratton, I., Doll, R., et al. (1987). Cancer near nuclear installations. *Nature, 329.*

Foucault, M. (1976). *La volonté de savoir. Histoire de la sexualité Tome 1.* Paris: Gallimard.

Franco, T., Druck, M. G. A Degradação do Trabalho e Os Riscos Industriais No Contexto da Globalização, Reestuturação Produtiva e das Políticas Neoliberais. In: Tânia Franco. (Org.). *Trabalho, riscos industriais e meio ambiente: Rumo ao desenvolvimento sustentável?* 01 ed. Salvador, Bahia: EDUFBA, 1997, pp. 15-32.

François, M. (1991). Le travail temporaire en milieu industriel. Incidences sur les conditions de travail et la santé des travailleurs. *Le Travail Humain, 54*(1).

Frigul, N. (1997). *Penser le rapport au travail pour comprendre le chômage. Une analyse de la construction sociale de la précarisation du travail et de la santé à partir d'une enquête menée auprès d'une population féminine en chômage de longue durée.* Thèse de doctorat en sciences sociales. Université René Descartes. Paris V.

Gardner, M. J., et al. (1990). Results of case-control study of leukaemia and lymphoma among young people near Sellafield nuclear plant in West Cumbria. *British Medical Journal, 300,* 423-429.

Gérôme, N. (1989). *Gestes et métiers de l'énergie Messidor.* Paris: Editions Sociales.

Gilbert, C. (dir.) (1998). *Actes du séminaire du programme "risques collectife et situations de crise."* CNRS, Grenoble.

Golac, M. (1997). Article Revue Française de socio.

Gorz, A. (1997). *Métamorphoses du travail. Quête du sens. Critique de la raison économique.* Paris: Galilée.

Gould, J. M., & Goldman, B. A. (1991). *Deadly deceit. Low-level radiation, high-level cover-up.* New York: Four Walls Eight Windows.

Greene, G. (2003). *The women who knew too much. Alice Stewart and the secrets of radiation.* The University of Michigan Press.

Grégoire, D. (2006). Radiations ionisantes. France: La situation des sous-traitants reste préoccupante (entretien avec Michel Lallier). *HSEA Newsletter* No. 29 mars.

Groupe de recherches historiques de la centrale de Cheviré. (1987). Une centrale et ses hommes. *Histoire de Cheviré 1953/1986.* Ed. ACL, Saint Sébastien.

Guignon, N., & Dr. Sandret, N. (2005). «Les expositions aux produits cancérogènes», *Premières Synthèses, 28*(1), juillet 2005, Dares.

Hancké, B. (1998). Trust or hierarchy? Changing relationships between large and small firms in France. *Small Business Economics, 11,* 237-252.

Haut-Comité à la Santé publique. (1998). *La progression de la précarité en France et ses effets sur la santé.* Paris, Editions E.N.S.P.

Hery, M., Gerber, J. M., Hubert, G., Hetcht, G., & Diebold, F. (1995). *Arrêt biennal d'un site de l'industrie chimique. Exposition aux produits chimiques des salariés d'entrerpises extérieures.* INRS Cahiers des notes documentaires No. 161, 4e trimestre.

Hery, M., Gerber, J. M., Hubert, G., Hetcht, G., Diebold, F., Honnert, B., et al. (1994). Exposure to metallic catalyst dust: Manufacturing and handling of catalysts in the chemical industry. *Annals of Occupational Hygiene, 38*(2), 119-135.

Hill, C. (1995). Risque de cancer autour des installations nucléaires. In: Problèmes liés aux effets des faibles doses des radiations ionisantes. *Rapport de l'Académie des Sciences,* no. 34, octobre.

Hirata, H. Dir. (1992). *Autour du "modèle" japonais. Automatisation, nouvelles formes d'organisation et de relations de travail.* Paris: L'Harmattan.

HSE. (1999). *Safety managment audit of British Energy Generation Limited and British Energy Generation (UK) Limited.* HM Nuclear Inspection Inspectorate.

Huez, D. (1996). La foundation du groupe "Paroles," et les nouveaux problémes psychologiques reneontrés dans la pratique de la médecine du travail. *Revue internationale de psychosociologie, 3*(5), 63-75.

Huez, D. (1998). *Précarité et radioprotection,* GNC, No. 346.

ICRP 26. (1997). Recommandations de la Commission Internationale de Radioprotection. Traduit par Mme A. Duchêne sous la direction de H. Jammet, CEA, IPSN Pergamont Press, Oxford.

ICRP 60. (1990). *The new international commission on radiological protection safety standards*, Oxford, novembre 1990.

IGSN. (2006). *La sûreté nucléaire et la radioprotection en 2005.* Rapport au président d'EDF, Paris.

IIPA/SSFI. (1997). *Evaluation findings and recommendations.* Rapport préparé à la demande de la direction d'Ontario Hydronuclear, Juillet 1997.

INSAG-4. (1991). *Culture de Sûreté.* Rapport du Groupe Consultatif International pour la Sûreté Nucléaire. Vienne, AIE.

IRSN. (2005a). *Risque de cancer associé aux faibles doses de rayonnements ionisants: une étude de cohorte rétrospective dans 15 pays.* Note de lecture, Fontenay-aux-roses.

IRSN. (2005b). *La radioprotection des travailleurs.* Bilan 2004, Rapport, Fontenay-aux-roses.

Janiaud, J. (1990). EDF et la main invisible. Paris: L'Harmattan.

Jobert, G. (1998). *La compétence à vivre. Contribution à une anthropologie de la reconnaissance au travail. Habilitation à diriger des recherches,* Université de Tours

Kergoat, J., Boutet, J., Jacot, H., & Linhart, D. (dir., 1998). *Le monde du travail.* Paris: La Découverte.

Kern, H., & Schuman, M. (1989). *La fin de la division du travail?: La rationnalisation dans la production industrielle.* Paris: la Maison des Sciences de l'Homme.

Kesminiene, A., & Cardis, E. (2007). l'épidémiologie de l'après-Tchernobyl. *Bulletin du cancer, 94,* 423-430, synthèse.

Kneale, G. et al. (1981, Summer). Hanford III, a cohort study of the cancer risks from radiation to workers at Hanford (1944 to 1977 deaths) by method of regression models in life-tables. *British Journal of Industrial Medicine, 38,* 56-66.

Kuhn, T. S. (1983). *La structure des révolutions scientifiques.* Paris: Flammarion, coll. "Champs."

Lacoster, A., & DSIN. (1998). Radioprotection et installations nucléaires de base, *Revue Contrôle,* ASN, 1998.

Lallemand, J., & Court R. (1997). Commentaire relatif à la lettre du professeur Clarke paru dans la Revue de Radioprotection de janvier 1997 (no. 4).

Lallier, M. (1995). *Sous-traitance. Le cas du nucléaire. Etude réalisée au Centre Nucléaire de Chinon.* Ed. CGT CNPE Chinon.

Lampercht, J-L. (1994). *ISO 9000, se préparer à la certification.* Paris: Afnor Editions.

Lanir, Z. (1986). *Fundamental surprise.* Eugene, OR: Decision Research.

Laville, A. (1984). *Temps de travail et santé. Séminaire franco-brésilien.* U.S.P., São Paulo, juillet 1984.

Laville, A. (1991). Plaidoyer pour le lampiste. *Santé et travail,* No. 2, septembre/octobre.

Laville, A., & Volkoff, S. (1993). *Age, santé, travail: Le déclin et la construction.* Actes du XXVIIe Congrès de la SELF, octobre.

Le Cohu, P. (1994). Le temps réduit indemnisé de longue durée. *Droit Social,* No. 2, février.

Le Déaut, J. Y. (1998). Rapport sur le système français de radioprotection, de contrôle et de sécurité nucléaire. Office parlementaire d'évaluation des choix scientifiques et techniques, Paris.

Le sexe du travail. (1984). *Structures familiales et systèmes productifs.* Grenoble, Presses Universitaires de Grenoble.

Lefaure, C., & Lochard, J. (1990). La dosimétrie des travailleurs des entreprises extérieures dans les centrales nucléaires. *Risque et Prévention,* No. 9, Novembre.

Lefaure, C. (1991). La part des entreprises extérieures dans le bilan de l'exposition professionnelle en France. *Radioprotection, 26*(4).

Les nouvelles formes de domination dans le travail; *Actes de la Recherche en Sciences Sociales,* No. 114, septembre 1996 et N° 115 décembre 1996.

Lilienfield, D. E. (1991, June). The sylence: The asbestos industry and early occupational cancer research—A case study. *American Journal of Public Health, 81*(6).

Lippel, K. (2004). Le travail atypique et la législation en matière de santé et sécurité du travail. In: *Développements récents en droit de la santé et sécurité au travail. Service de la formation permanente du Barreau du Québec* (Vol. 201, Editions Yvon Blais, pp. 307-382).

Llory, M. (1996). *Accidents industriels: Le coût du travail.* Chinon: L'Harmattan.

Malye, F. (1996). *Amiante. Le dossier de l'air contaminé.* Paris: Le pré aux Clercs.

Mancuso, T. F, Stewart A., & Kneale G. (1977, November). Radiation exposures of Hanford workers dying from cancer and other causes. *Health Physics Journal, 33*(5), 369-384.

Mancuso, T. F. et al. (1978). *A reanalysis of data relating to the Hanford study of the cancer risks of radiation workers.* International Atomic Agency Symposium Proceedings on the Late Biological Effects of Ionizing Radiation, Vienna, Austria, IAEA-SM-224/510.

Mancuso, T. F. et al. (1981, March). *Hanford IIIb, delayed effects of small doses of radiation delivered at slow dose rates.* Proceedings of a Symposium on Industrial Cancers, Cold Spring Harbor, Banberry Center, Long Island, New York.

Marquié, J. C., Volkoff, S., & Paumès, D. (1996). *Le travail au fil de l'âge.* Toulouse, ed. Octares.

Martin, G. J. (1995). *Précaution et évolution du droit.* Paris: Dalloz.

Mauss, M. (1923/24). Essai sur le don. Forme et raison de l'échange dans les sociétés archaïques. In: M. Mauss (1985) *Sociologie et anthropologie.* Paris: PUF (9e édition).

Mayhew, C., Quinlan, M., & Bennet, L. (1996). *The effects of subcontracting/outsourcing on occupational health and safety.* Industrial Relations Research Center, University of New South Wales, Studies in Australian Industrial Relations, No. 38.

Mayoux, M. (1990). *Aspect juridique du rôle et de la responsabilité des différents intervenants. Colloque sur la Surveillance de l'Exposition Individuelle, Société Francaise de Radioprotection,* Paris 17-18 octobre.

Mendel, G. (1997). L'institution au double péril de l'organisation et de l'inconscient: Perspectives sociopsychanalytiques (1971-1997) In: *Psychanalyse et organisation, Revue internationale de psychosociologie, IV*(6/7).

Meynaud, H. Y. dir. (1996). *Les sciences sociales et l'entreprises. Cinquante ans de recherche à EDF.* Paris: La Découverte.

Mispelblom, F. (1995). *Au-delà de la qualité. Démarches qualité, conditions de travail et politiques du bonheur.* Parsi: Syros.

Molinier, P., Doniol-Shaw, G., Dessors, D., & Huez, D. (1996). Sous-traitance et exclusion sociale. *Pratiques Psychologiques 1996-1.*

Morgan, K. Z. (1980). The hazards of low-level radiation. *Encyclopaedia Britannica,* 1980 Edition.

Morin, M. L. (1994). *La sous-traitance et les relations salariales. Rapport de recherche pour le Commissariat Général au plan.* CEJEE, Toulouse.

Mossman, K. L. (1998, March). The linear no-threshold debate: Where do we go from here? *Medical Physics, 25*(3).

Pages, P., & Hubert, P. H. (1994). *Analyse de l'exposition des travailleurs extérieurs aux rayonnements ionisants dans l'industrie nucléaire,* IPSN, Note SEGR/LSEES-93/ No. 24 Révision 1-Mai 1994.

Parker, L. et al. (1993). Geographical distribution of preconceptional radiation doses to fathers employed at Sellafield nuclear installation, West Cumbria. *British Medical Journal, 307,* 966-971.

Paugam, S. (1996). *Dir. L'exclusion: l'état des savoirs.* Paris: La Découverte.

Pereira, V., Remoiville, A., & Trinquet, P. (1999). *Sous-traitance sur sites industriels: Évaluation des risques professionnels.* Rapport de recherche, Marseille, APRIT, ARESI-BTP.

Perrow, C. (1984). *Normal accidents. Living with high-risk technologies.* New York: Basic Books.

Pesteil, J. M. (1997). *Optimisation de la maintenance conditionnelle.* EDF-CNPE Chinon.

Petrella, R. (1996). *Le biem commun. Eloge de la solidarité.* Éditions Labor.

Pille, P. (1991). Etude de l'exposition externe des travailleurs non EDF intervenant dans les centrales électronucléaires: Essai de quantification pour certains travaux. *Revue de Médecine du Travail, XVIII*(1), janvier-février.

Piore, M. J., & Sabel, C. F. (1984). *The second industrial divide: Possibilities for prosperity.* New York: Basic Books.

Pobel, D., & Viel, J. F. (1997, January). Case-control study of leukemia among young people near La Hague nuclear reprocessing plant: The environmental hipothesis revisited. *British Medical Journal, 314*(1).

Polack, M. (1993). *Une identité blessée.* ed. A. Metaillé, Paris.

Puiseux, L. (1986). *Crépuscule des atomes.* Paris: Hachette.

Querrien, M. (1991). Lutter contre les accidents de chantier, c'est possible. *Santé et travail,* No. 2 septembre-octobre.

Quinlan, M., & Mayhew, C. (1999). Precarious employment and workers' compensation. *International Journal of Law and Psychiatry, 22*(5-6), 491-520.

Quinlan, M., & Mayhew, C. (2000). Precarious employment, work re-organisation and the facturing of OHS management. In: K. Frick, P. L. Jensen, M. Quinlan, & T. Wilthagen (Eds.), *Systematic occupational health and safety management. Perspectives on an international development* (pp. 175-198). Oxford: Pergamon/ Arbetslivsinstitutet, Elsevier Science.

Quinlan, M., Mayhew, C., & Bohle, P. (2001). The global expansion of precarious employment, work disorganisation, and consequences for occupational health: A review of recent research. *International Journal of Health Services, 31*(2/3).

Rapport Noc. (1990). *Rapport sur l'amélioration de la sûreté nucléaire en exploitation.* SPT/EDF.

Reason, J. (1990). *Human error.* Cambridge: Cambridge University Press.

Réseau SCOP93. (2005). A multidisciplinary network about occupational cancer in a Paris suburb, Seine Saint Denis (France), First results of a pro-active research. *International Journal of Occupational and Environnemental Health, 11,* 263-275.

Rettenbach, B. (1978). Droit du travail et restructurations productives. Les nouveaux contrats de travail. Paris, la Documentation Française.

RGN-Actualités. (1992). *Le carnet d'accès.* Une gestion nouvelle pour la sécurité des travailleurs RGN N°5 Septembre-octobre.

Ritz, B., Morgenstern, H., Froines, J., & Bambi, B. Y. (1999). Effects of exposure to external ionizing radiation on cancer mortality in nuclear workers monitored for tadiation at Rocketdyne/Atomics International. *American Journal of Industrial Medicine, 35,* 21-31.

Roche, P. (1987). *Une santé d'acier.* Paris: Ed. du CNRS.

Rot, G. (1998). Autocontrôle, traçabilité, responsabilité. *Sociologie du Travail, XL,* 1/98

Salminen, S., Saari, J., Saarela, K. L., & Rasanen, T. (1993). Scandinavian Journal organizational factors influencing serious occupational accidents. *Work Environmental Health, 19,* 352-357.

Seillan, H., & Morvan, J. (2005). Risques de la sous-traitance. 2e Edition revue et complétée. *Les cahiers de Préventique,* no. 4, Editions Préventique, Bordeaux.

Souffrances et précarités au travail. Paroles de médecins du travail. (1994). Paris: Syros.

Stewart, A., & Kneale, G. (1982). Delayed effects of A-Bom Radiation: A review of recent mortality rates and risk estimates for five-years survivors. *Journal of Epidemiology and Community Health, 36*(2), 80-86.

Stewart, A., & Kneale, G. (1993). A-Bomb survivors: Further evidence of late effects of early deaths. *Health Physics, 64,* 467-472.

Stewart, A. et al. (1980, June). Hanford IIb, The Hanford Data—A reply to recent criticisms. *Ambio 9,* 66-73.

Supiot, A. (1994). *Critique du Droit du Travail.* Paris: PUF.

Surault, P. (1997). Suicide et milieu social. In: J. J. Chavagnat & R. Franc (Coord.), *Suicide et vie professionnelle: Les risques du métier,* GEPS, Stersup Ed., Toulouse.

Tanaka, Y. (1997). Nuclear power plant gypsies in high-tech society. In: J. Moore (Ed.), *The other Japan. Conflict, compromise and resistance since 1945.* New York: M. E. Sharpe.

Teiger, C. (1981). Travailleurs de nuit permanents, rythmes circadiens et mortalité. *Le Travail Humain, 44*(1).

Thébaud-Mony, A. (1984). *Sous-traitance, division du travail, division des risques et santé. Enquête auprès des travailleurs extérieurs d'un site industriel en région parisienne. Séminaire franco-brésilien.* Université de Sao Paulo, Juillet.

Thébaud-Mony, A. (1990). *L'envers des sociétés industrielles. Approche comparative franco-brésilienne.* Paris: L'Harmattan.

Thébaud-Mony, A. (1991). *La reconnaissance des maladies professionnelles. Travail et Emploi, 1er Numéro hors-série.* Paris: la Documentation Française.

Thébaud-Mony, A. (1994, Winter). Compensation of occupational diseases in France. *New Solutions. Journal of Environmental and Occupational Health Policy, 4*(2).

Thébaud-Mony, A. (1994a). La précarité moderne. *Politis-La Revue, 7,* Avril-Mai-Juin.

Thébaud-Mony, A. (1994b). Vers une société précaire. *Santé et Travail,* No. 8, mars/avril.

Thébaud-Mony, A. (1995). *Sous traitance, précarisation du travail, risques professionnels et santé. Rapport de recherche.* Kremlin Bicêtre: INSERM U292.

Thébaud-Mony, A. (1997a). La santé au travail: Enjeu et instrument de la précarisation sociale. In: B. Appay, & A. Thébaud-Mony (dir). *Précarisation sociale, travail, santé.* Paris: IRESCO.

Thébaud-Mony, A. (1997b). Rationnalité instrumentale et santé au travail: Le cas de l'industrie nucléaire. *Revue Internationale de Psychosociologie, III*(8).

Thébaud-Mony, A., Rondeau Du Noyer, C., Huez, D., Brener, F., Forest, H., Géraud, G., et al. (1992). *Salariés des entreprises extérieures intervenant dans les installations nucléaires de base. Enquête sur leur suivi médico-règlementaire.* Documents pour les médecins du travail, INRS, No. 51.

Thomas, L. V. (1993). *Mélanges thanatiques. Nouvelles Etudes Anthropologiques.* Paris: L'Harmattan.

Tubiana, M., & Aurengo, A. (2005). *La relation dose-effet et l'estimation des effets cancérogènes des faibles doses de rayonnements ionisants. Rapport d'un groupe de travail mixte de l'Académie de Médecine et de l'Académie des Sciences,* Paris.

Tubiana, M., & Aurengo, A. (2006). Point de vue d'experts: l'Académie française de Médecine et de l'Académie Française des Sciences. In: Controverse: Les faibles doses de radiations ionisantes sont-elles carcinogéniques? *Bulletin Epidémiologique Hebdomadaire,* No. 15-16/2006.

Turshen, M. (1989). *The politics of public health.* New Brunswick, NJ: Rutgers University Press.

Urquhart, J. (1991). Case control study of leukaemia and non-Hodgkin's lymphoma in children in Caithness near the Dounreay nuclear installation. *British Medical Journal, 27*(11), 1486-1490.

Urquhart, J. (1993, January 9). Incidence of leukaemia and non-Hodgkin's lymphoma in the vicinity of nuclear sites in Scotland, 1968-1993. *Journal of Epidemiology and Community Health, 306,* 89-94.

Vacarie, I. (1982). Stratégie d'emploi des grandes entreprises et droit du travail. In: *Colloque de Dourdan.* Paris: Maspero.

Vacarie, I., & Supiot, A. (1993). Santé, sécurité et libre circulation des marchandises (règles juridiques et normes techniques). *Droit Social* No. 1.

Vandevyver, C. (1984). Risques d'accidents liés à l'intervention de personnel d'entreprises extérieures. *Cahiers de Notes documentaires,* INRS, No. 115, 2e trimestre.

Viel, J. F. et al. (1993). Childhood leukemia incidence in the vicinity of La Hague nuclear-waste reprocessing facility (France). *Cancer Causes and Control, 4.*

Viel, J. F. (1998). *La santé publique atomisée: Radioactivité et leucémies: Les leçons de La Hague.* Paris: La Découverte.

Vogel, L. (1994). *L'organisation de la prévention sur les lieux de travail. Un premier bilan de la Directive-cadre de 1989.* Bureau Technique Syndical Européen pour la Santé et la Sécurité, Bruxelles.

Vogel, L. (1997). *Droit du travail et précarisation des droits dans les états de l'Union Européenne.* In: B. Appay & A. Thébaud-Mony (dir). Précarisation sociale, travail et santé. Paris, IRESCO.

Walraff, G. (1985). *Ganz unten.* Cologne: Verlag Kiepenheuer & Witsch.

Walraff, G. (1986). *Tête de Turc* (Traduit de l'allemand par A. Brossat et K. Schuffels). Paris: La Découverte.

Wig, S., Richardson, D., & Stewart, A. (1999). The relevance of occupational epidemiology to radiation protection standard. *New Solutions, 9*(2), 133-151.

Wing, S. (1998). Job factors, radiation and cancer mortality at Oak Ridge National Laboratory: Follow-up through 1984. *American Journal of Industrial Medicine, 23,* 265-279.

Woods, D. D. (1982). *Operator decision behavior during the steam generator tube rupture at the ginna nuclear power station*, Research Report 82-1c57-CONRM-R2. Pittsburg, PA: Westinghouse R&D Center.

Zarifian, P. (1996). *Travail et communication. Essai sociologique sur le travail dans la grande entreprise industrielle.* Paris: PUF.

Zerbib, J. C. (1997a). Faibles doses de rayonnements. Un excès de cancers statistiquement significatifs. *Santé et travail,* no. 19 avril 1997.

Zerbib, J. C. (1997b). Will new ionizing radiations directive increase workers and public safety? *TUTB Newsletter,* No. 6, June 1997.

Zerbib, J. C., & Forest, H. (1991). Le suivi dosimétrique des travailleurs des "entreprises extérieures" intervenant dans les installations nucléaires. *Texte syndical,* octobre 1999.

Glossary

Contamination: refers to the presence of radioactive particles or dust. Inhalation or ingestion of such particles or dust leads to "internal contamination"; "external contamination" refers to contact with skin or hair.
Contractor: company that subcontracts certain tasks or productive functions to other companies.
Controlled access zones: zones in nuclear power plants where there is a risk of irradiation or radiation contamination. These zones are subject to constant radioactivity monitoring. Persons entering them must wear an instrument for measuring radioactivity.
DATR (Directement Affecté aux Travaux sous Rayonnements): initials used in France to designate nuclear industry workers who have taken training programs in radiation protection and are subject to the special medical surveillance provided for by the Code du Travail for workers performing tasks that expose them to ionizing radiation.
Dose: refers in nuclear industry language not to the actual measure of radiation emitted by a radioactive source but a complex measure of possible biological effects that the experts call "dose equivalent."
Millisievert (mSv): see Rem.
Nuclear servitude: set of services provided to the nuclear industry by service companies, including radioactive decontamination, installing and dismantling of scaffolding and airlocks, lagging/de-lagging, and various radiation protection tasks.
Outside workers: all employees of subcontractor companies and temporary agencies who perform maintenance tasks on user or contractor sites.
Radiation protection: in labor law, the set of provisions aimed at protecting workers from ionizing radiation.
Rem, millisievert: units for measuring radioactivity: 1 rem = 10 mSv. The current regulation unit is the mSv, but in ordinary language, workers continue to use "rem" and "millirem."
Safety: **Worker Safety**: the set of provisions adopted to protect workers' health. In this text, the term refers especially to radiation protection. **General Safety**: the set of provisions aimed at maximally reducing risk of nuclear accident and radiation dissemination.

Shutdown: period when nuclear reactors are brought to a halt so that inspection and maintenance work may be done.

Subcontracting: defined by AFNOR (norm NFX 50 300N of November 1987) as "all activities involved in realizing one or more design, development, fabrication, operation or maintenance operation(s) for the product in question for a given production cycle that a given company, called the contractor, entrusts to another company, called the subcontractor. The subcontractor must comply exactly with the technical orders and specifications determined from first to last by the contractor."

Sub-subcontracting: the practice whereby a contractor concludes a business agreement with one or several companies that have themselves concluded business agreements with other companies.

Task vs. Activity: distinction made by ergonomics: the task is prescribed work; the activity is the real work that workers perform. Discrepancies exist between prescribed and real work.

To be "on the peg": have reached the maximum individual dose permitted by the regulations.

To be "out to pasture": expression used by outside company workers operating in nuclear power plants; refers to being laid off for having reached maximum regulation individual dose.

Index